CAMARO
MUSCLE PORTFOLIO
——— 1967-1973 ———

Compiled By R.M. Clarke

ISBN 1 85520 1453

BROOKLANDS BOOKS LTD.
P.O. BOX 146, COBHAM,
SURREY, KT11 1LG. UK

A-CAM67MP

Printed in Hong Kong

Brooklands Books

MOTORING

B.B. ROAD TEST SERIES

Abarth Gold Portfolio 1950-1971
AC Ace & Aceca 1953-1983
Alfa Romeo Giulietta Gold Portfolio 1954-1965
Alfa Romeo Giulia Coupés 1963-1976
Alfa Romeo Giulia Coupés Gold Port. 1963-1976
Alfa Romeo Spider 1966-1990
Alfa Romeo Spider Gold Portfolio 1966-1991
Alfa Romeo Alfasud 1972-1984
Alfa Romeo Alfetta Gold Portfolio 1972-1987
Alfa Romeo Alfetta GTV6 1980-1986
Allard Gold Portfolio 1937-1959
Alvis Gold Portfolio 1919-1967
AMX & Javelin Muscle Portfolio 1968-1974
Armstrong Siddeley Gold Portfolio 1945-1960
Aston Martin Gold Portfolio 1948-1971
Aston Martin Gold Portfolio 1972-1985
Aston Martin Gold Portfolio 1985-1995
Audi Quattro Gold Portfolio 1980-1991
Austin A30 & A35 1951-1962
Austin-Healey 100 & 100/6 Gold Port. 1952-1959
Austin-Healey 3000 Ultimate Portfolio 1959-1967
Austin-Healey Sprite Gold Portfolio 1958-1971
BMW 6 & 8 Cyl. Cars Limited Edition 1935-1960
BMW 1600 Collection No. 1 1966-1981
BMW 2002 Gold Portfolio 1968-1976
BMW 6 Cylinder Coupés & Saloons Gold P. 1969-1976
BMW 316, 318, 320 (4 cyl.) Gold Port. 1975-1990
BMW 320, 323, 325 (6 cyl.) Gold Port. 1977-1990
BMW 3 Series Gold Portfolio 1991-1997
BMW 5 Series Gold Portfolio 1981-1987
BMW 5 Series Gold Portfolio 1988-1995
BMW 6 Series Gold Portfolio 1976-1989
BMW 7 Series Performance Portfolio 1977-1986
BMW Alpina Performance Portfolio 1967-1987
BMW Alpina Performance Portfolio 1988-1998
BMW M Series Gold Portfolio 1976-1997
BMW Z3 & Z3M Limited Edition
Borgward Isabella Limited Edition
Bricklin Gold Portfolio 1974-1975
Bristol Cars Gold Portfolio 1946-1992
Buick Automobiles 1947-1960
Buick Muscle Cars 1965-1970
Cadillac Allanté 1986-1993
Cadillac Automobiles 1949-1959
Cadillac Automobiles 1960-1969
Checker Limited Edition
Chevrolet 1955-1957
Impala & SS Muscle Portfolio 1958-1972
Corvair Performance Portfolio 1959-1969
El Camino & SS Muscle Portfolio 1959-1987
Chevy II & Nova SS Muscle Portfolio 1962-1974
Chevelle & SS Muscle Portfolio 1964-1972
Caprice Limited Edition 1965-1976
Chevrolet Muscle Cars 1966-1971
Chevy Blazer 1969-1981
Camaro Muscle Portfolio 1967-1973
Chevrolet Camaro & Z-28 1973-1981
High Performance Camaros 1982-1988
Chevrolet Corvette Gold Portfolio 1953-1962
Chevrolet Corvette Sting Ray Gold Port. 1963-1967
Chevrolet Corvette Gold Portfolio 1968-1977
High Performance Corvettes 1983-1989
Chrysler 300 Gold Portfolio 1955-1970
Imperial Limited Edition 1955-1970
Valiant 1960-1962
Citroen Traction Avant Gold Portfolio 1934-1957
Citroen 2CV Gold Portfolio 1948-1989
Citroen DS & ID 1955-1975
Citroen DS & ID Gold Portfolio 1955-1975
Citroen SM 1970-1975
Cobras & Replicas 1962-1983
Shelby Cobra Gold Portfolio 1962-1969
Cobras & Cobra Replicas Gold Portfolio 1962-1989
Crosley & Crosley Specials Limited Edition
Cunningham Automobiles 1951-1955
Daimler SP250 Sports & V-8 250 Saloon
Ultimate Portfolio 1959-1969
Datsun Roadsters 1962-1971
Datsun 240Z & 260Z Gold Portfolio 1970-1978
Datsun 280Z & ZX 1975-1983
DeLorean Gold Portfolio 1977-1995
De Soto Limited Edition 1952-1960
Charger Muscle Portfolio 1966-1974
Dodge Viper Performance Portfolio 1990-1998
ERA Gold Portfolio 1934-1994
Excalibur Collection No.1 1952-1981
Facel Vega 1954-1964
Ferrari Limited Edition 1947-1957
Ferrari Limited Edition 1958-1963
Ferrari Dino 1965-1974
Ferrari Dino 308 & Mondial Gold Portfolio 1974-1985
Ferrari 328 348 Mondial Gold Portfolio 1986-1994
Fiat 500 Gold Portfolio 1936-1972
Fiat 600 & 850 Gold Portfolio 1955-1972
Fiat Pininfarina 124 & 2000 Spider 1968-1985
Fiat X1/9 Gold Portfolio 1973-1989
Fiat Abarth Performance Portfolio 1972-1987
Ford Consul, Zephyr, Zodiac Mk. I & II 1950-1962
Ford Zephyr, Zodiac, Executive Mk. III & IV 1962-1971
Ford Cortina 1600E & GT 1967-1970
High Performance Capris Gold Portfolio 1969-1987
Capri Muscle Portfolio 1974-1987
High Performance Fiestas 1979-1991
Ford Escort RS & Mexico Limited Edition 1970-1979
High Performance Escorts Mk. I 1968-1974
High Performance Escorts Mk. II 1975-1980
High Performance Escorts 1980-1985
High Performance Escorts 1985-1990
High Perf. Sierras & Merkurs Gold Port. 1983-1990
Ford Automobiles 1949-1959
Ford Fairlane Performance Portfolio 1955-1970
Ford Ranchero Muscle Portfolio 1957-1979
Edsel Limited Edition 1957-1960
Falcon Performance Portfolio 1960-1970
Ford Galaxie & LTD Limited Edition 1960-1973
Ford Thunderbird 1955-1957
Ford Thunderbird 1958-1963
Ford GT40 Gold Portfolio 1964-1987
Ford Torino Limited Edition 1968-1974
Ford Bronco 4x4 Performance Portfolio 1966-1977
Ford Bronco 1978-1988
Goggomobil Limited Edition
Holden 1948-1962
Honda S500 • S600 • S800 Limited Edition 1962-1970
Honda CRX 1983-1987
Hudson Limited Edition 1946-1957
International Scout Gold Portfolio 1961-1980
Isetta Gold Portfolio 1953-1964
ISO & Bizzarrini Gold Portfolio 1962-1974
Jaguar and SS Gold Portfolio 1931-1951
Jaguar C-Type & D-Type Gold Portfolio 1951-1960
Jaguar XK120, 140, 150 Gold Portfolio 1948-1960
Jaguar Mk. VII, VIII, IX, X, 420 Gold Port. 1950-1970
Jaguar Mk. 1 & Mk. 2 Gold Portfolio 1959-1969
Jaguar E-Type Gold Portfolio 1961-1971
Jaguar E-Type V-12 1971-1975
Jaguar S-Type & 420 Limited Edition
Jaguar XJ12, XJ5.3, V12 Gold Portfolio 1972-1990
Jaguar XJ6 Series I & II Gold Portfolio 1968-1979
Jaguar XJ6 Series III Perf. Portfolio 1979-1986
Jaguar XJ6 Gold Portfolio 1986-1994
Jaguar XJS Gold Portfolio 1975-1988
Jaguar XJS Gold Portfolio 1988-1995
Jaguar XK8 Limited Edition
Jeep CJ5 & CJ6 1960-1976
Jeep CJ5 & CJ7 4x4 Perf. Portfolio 1976-1986
Jeep Wagoneer Performance Portfolio 1963-1991
Jeep J-Series Pickups 1970-1982
Jeep Wrangler 4x4 1986-1998
Jeep Cherokee & Grand Cherokee 4x4
Performance Portfolio 1992-1998
Jensen Interceptor Gold Portfolio 1966-1986
Jensen - Healey Limited Edition 1972-1976
Kaiser - Frazer Limited Edition 1946-1955
Lagonda Gold Portfolio 1919-1964
Lancia Aurelia & Flaminia Gold Portfolio 1950-1970
Lancia Fulvia Gold Portfolio 1963-1976
Lancia Beta Gold Portfolio 1972-1984
Lancia Delta Gold Portfolio 1979-1994
Lancia Stratos 1972-1985
Land Rover Series I 1948-1958
Land Rover Series II & IIa 1958-1971
Land Rover Series III 4x4 Perf. Portfolio 1971-1985
Land Rover 90 110 Defender Gold Portfolio 1983-1994
Land Rover Discovery 1989-1994
Land Rover Story Part One 1948-1971
Fifty Years of Selling Land Rover
Lincoln Gold Portfolio 1949-1960
Lincoln Continental 1961-1969
Lincoln Continental 1969-1976
Lotus Sports Racers Gold Portfolio 1953-1965
Lotus Seven Gold Portfolio 1957-1973
Lotus Caterham Seven Gold Portfolio 1974-1995
Lotus Elan Gold Portfolio 1962-1974
Lotus Elan & SE 1989-1992
Lotus Europa Gold Portfolio 1966-1975
Lotus Elite & Eclat 1974-1982
Marcos Coupés & Spyders Gold Portfolio 1960-1997
Matra Limited Edition 1965-1983
Mazda Miata MX-5 Performance Portfolio 1989-1997
Mazda RX-7 Gold Portfolio 1978-1991
McLaren F1 Sportscar Limited Edition
Mercedes 190 & 300 SL 1954-1963
Mercedes G-Wagen 1981-1994
Mercedes S & 600 1965-1972
Mercedes S Class 1972-1979
Mercedes 230 • 250 • 280SL Gold Portfolio 1963-1971
Mercedes SLs & SLCs Gold Portfolio 1971-1989
Mercedes SLs Performance Portfolio 1989-1994
Mercury Limited Edition 1947-1959
Mercury Comet & Cyclone Limited Edition 1960-1970
Mercury Muscle Cars 1966-1971
Cougar Limited Edition 1967-1973
Messerschmitt Gold Portfolio 1954-1964
MG Gold Portfolio 1929-1939
MG TA & TC Gold Portfolio 1936-1949
MG TD & TF Gold Portfolio 1949-1955
MGA & Twin Cam Gold Portfolio 1955-1962
MG Midget Gold Portfolio 1961-1979
MGB Roadsters 1962-1980
MGB MGC & V8 Gold Portfolio 1962-1980
MGB GT 1965-1980
MGC & MGB GT V8 Limited Edition
MG Y-Type & Magnette ZA/ZB Limited Edition
MGF Limited Edition
Mini Gold Portfolio 1959-1969
Mini Gold Portfolio 1969-1980
Mini Gold Portfolio 1981-1997
High Performance Minis Gold Portfolio 1960-1973
Mini Cooper Gold Portfolio 1961-1971
Mini Moke Gold Portfolio 1964-1994
Morgan Three-Wheeler Gold Portfolio 1910-1952
Morgan Plus 4 & Four 4 Gold Portfolio 1936-1967
Morris Minor Collection No. 1 1948-1980
Shelby Mustang Muscle Portfolio 1965-1970
High Performance Mustang IIs 1974-1978
Mustang 5.0L Muscle Portfolio 1982-1993
Nash & Nash-Healey Limited Edition 1949-1957
Nash-Austin Metropolitan Gold Portfolio 1954-1962
NSU Ro80 Limited Edition
Oldsmobile Automobiles 1955-1963
Oldsmobile Muscle Portfolio 1964-1971
Cutlass & 4-4-2 Muscle Portfolio 1964-1974
Oldsmobile Toronado 1966-1978
Opel GT Gold Portfolio 1968-1973
Opel Manta Limited Edition 1970-1975
Packard Gold Portfolio 1946-1958
Pantera Gold Portfolio 1970-1989
Panther Gold Portfolio 1972-1990
Barracuda Muscle Portfolio 1964-1974
Pontiac Limited Edition 1949-1960
Pontiac Tempest & GTO 1961-1965
GTO Muscle Portfolio 1964-1974
Firebird & Trans-Am Muscle Portfolio 1967-1972
Firebird & Trans-Am Muscle Portfolio 1973-1981
High Performance Firebirds 1982-1988
Pontiac Fiero 1984-1988
Porsche 356 Gold Portfolio 1953-1965
Porsche 912 Limited Edition
Porsche 911 1965-1969
Porsche 911 1970-1972
Porsche 911 1973-1977
Porsche 911 SC & Turbo Gold Portfolio 1978-1983
Porsche 911 Carrera & Turbo Gold Port. 1984-1989
Porsche 911 Gold Portfolio 1990-1997
Porsche 924 Gold Portfolio 1975-1988
Porsche 928 Performance Portfolio 1977-1994
Porsche 928 Head to Head - Comparison Tests
Porsche 944 Gold Portfolio 1981-1991
Porsche 968 Limited Edition
Porsche Boxster Limited Edition
Railton & Brough Superior Gold Portfolio 1933-1950
Range Rover Gold Portfolio 1970-1985
Range Rover Gold Portfolio 1986-1995
Reliant Scimitar 1964-1986
Renault Alpine Gold Portfolio 1958-1994
Riley Gold Portfolio 1924-1939
R. R. Silver Cloud & Bentley 'S' Series Gold P. 1955-65
Rolls Royce Silver Shadow Ultimate Portfolio 1965-80
Rolls Royce & Bentley Gold Portfolio 1980-1989
Rolls Royce & Bentley Limited Edition 1990-1997
Rover P4 1949-1959
Rover 3 & 3.5 Litre Gold Portfolio 1958-1973
Rover 2000 & 2200 1963-1977
Rover 3500 & Vitesse 1976-1986
Saab Sonett Collection No.1 1966-1974
Saab Turbo 1976-1983
Studebaker Gold Portfolio 1947-1966
Studebaker Hawks & Larks 1956-1963
Avanti 1962-1990
Suzuki SJ Gold Portfolio 1971-1997
Vitara, Sidekick & Geo Tracker Perf. Port. 1988-1997
Sunbeam Tiger & Alpine Gold Portfolio 1959-1967
Toyota Land Cruiser Gold Portfolio 1956-1987
Toyota Land Cruiser 1988-1997
Toyota MR2 Gold Portfolio 1984-1997
Triumph TR2 & TR3 Gold Portfolio 1952-1961
Triumph TR4, TR5, TR250 1961-1968
Triumph TR6 Gold Portfolio 1969-1976
Triumph TR7 & TR8 Gold Portfolio 1975-1982
Triumph Herald 1959-1971
Triumph Vitesse 1962-1971
Triumph Spitfire Gold Portfolio 1962-1980
Triumph 2000, 2.5, 2500 1963-1977
Triumph GT6 Gold Portfolio 1966-1974
Triumph Stag Gold Portfolio 1970-1977
Triumph Dolomite Sprint Limited Edition
TVR Gold Portfolio 1959-1986
TVR Performance Portfolio 1986-1994
VW Beetle Gold Portfolio 1935-1967
VW Beetle Gold Portfolio 1968-1991
VW Beetle Collection No.1 1970-1982
VW Karmann Ghia 1955-1982
VW Bus, Camper, Van 1954-1967
VW Bus, Camper, Van 1968-1979
VW Bus, Camper, Van 1979-1989
VW Scirocco 1974-1981
VW Golf GTI 1976-1986
Volvo PV444 & PV544 1945-1965
Volvo Amazon-120 Ultimate Portfolio 1956-1970
Volvo 1800 Gold Portfolio 1960-1973
Volvo 140 & 160 Series Gold Portfolio 1966-1975
Forty Years of Selling Volvo
Westfield Limited Edition

B.B. *ROAD & TRACK* SERIES

Road & Track on Alfa Romeo 1964-1970
Road & Track on Alfa Romeo 1971-1976
Road & Track on Aston Martin 1962-1990
R & T on Auburn Cord and Duesenburg 1952-84
Road & Track on Audi & Auto Union 1952-1980
Road & Track on Audi & Auto Union 1980-1986
Road & Track on Austin Healey 1953-1970
Road & Track on BMW Cars 1966-1974
Road & Track on BMW Cars 1975-1978
Road & Track on BMW Cars 1979-1983
R & T on Cobra, Shelby & Ford GT40 1962-1992
Road & Track on Corvette 1953-1967
Road & Track on Corvette 1968-1982
Road & Track on Corvette 1982-1986
Road & Track on Corvette 1986-1990
Road & Track on Ferrari 1975-1981
Road & Track on Ferrari 1981-1984
Road & Track on Ferrari 1984-1988
Road & Track on Fiat Sports Cars 1968-1987
Road & Track on Jaguar 1950-1960
Road & Track on Jaguar 1961-1968
Road & Track on Jaguar 1968-1974
Road & Track on Jaguar 1974-1982
Road & Track on Jaguar 1983-1989
Road & Track on Lamborghini 1964-1985
Road & Track on Lotus 1972-1983
R & T on Mazda RX-7 & MX-5 Miata 1986-1991
Road & Track on Mercedes 1952-1962
Road & Track on Mercedes 1963-1970
Road & Track on Mercedes 1971-1979
Road & Track on Mercedes 1980-1987
Road & Track on MG Sports Cars 1949-1961
Road & Track on MG Sports Cars 1962-1980
R & T on Nissan 300-ZX & Turbo 1984-1989
Road & Track on Pontiac 1960-1983
Road & Track on Porsche 1951-1967
Road & Track on Porsche 1968-1971
Road & Track on Porsche 1972-1975
Road & Track on Porsche 1975-1978
Road & Track on Porsche 1979-1982
Road & Track on Porsche 1985-1988
R & T on Rolls Royce & Bentley 1950-1965
R & T on Rolls Royce & Bentley 1966-1984
Road & Track on Saab 1972-1992
R & T on Toyota Sports & GT Cars 1966-1984
R & T on Triumph Sports Cars 1953-1967
R & T on Triumph Sports Cars 1967-1974
R & T on Triumph Sports Cars 1974-1982
Road & Track on Volkswagen 1951-1968
Road & Track on Volkswagen 1968-1978
Road & Track on Volkswagen 1978-1985
Road & Track on Volvo 1957-1974
Road & Track on Volvo 1977-1994
Road & Track - Henry Manney at Large & Abroad
Road & Track - Peter Egan's "Side Glances"
Road & Track - Peter Egan "At Large"
Road & Track - Best of PS

B.B. *PRACTICAL CLASSICS* SERIES

PC on Austin A40 Restoration
PC on Land Rover Restoration
PC on Metalworking in Restoration
PC on Midget/Sprite Restoration
PC on MGB Restoration
PC on Sunbeam Rapier Restoration
PC on Triumph Herald/Vitesse

RACING

The Carrera Panamericana Mexico - 1950-1954
Le Mans - The Bentley & Alfa Years - 1923-1939
Le Mans - The Jaguar Years - 1949-1957
Le Mans - The Ferrari Years - 1958-1965
Le Mans - The Ford & Matra Years - 1966-1974
Le Mans - The Porsche Years - 1975-1982
Le Mans - The Porsche & Jaguar Years - 1983-91
Mille Miglia - The Alfa & Ferrari Years - 1927-1951
Mille Miglia - The Ferrari & Mercedes Years - 1952-57
Targa Florio - The Ferrari & Lancia Years - 1948-1954
Targa Florio - The Porsche & Ferrari Years - 1955-1964
Targa Florio - The Porsche Years - 1965-1973

A COMPREHENSIVE GUIDE

BMW 2002

B.B. *CAR AND DRIVER* SERIES

Car and Driver on BMW 1955-1977
Car and Driver on Corvette 1978-1982
Car and Driver on Corvette 1983-1988
C and D on Datsun Z 1600 & 2000 1966-1984
Car and Driver on Ferrari 1955-1962
Car and Driver on Ferrari 1963-1975
Car and Driver on Ferrari 1976-1983
Car and Driver on Mopar 1956-1967
Car and Driver on Mustang 1964-1972
Car and Driver on Pontiac 1961-1975
Car and Driver on Porsche 1955-1962
Car and Driver on Porsche 1963-1970
Car and Driver on Porsche 1970-1976
Car and Driver on Porsche 1977-1981
Car and Driver on Porsche 1982-1986
Car and Driver on Volvo 1955-1986

B.B. *HOT ROD* 'ENGINE' SERIES

Chevy 265 & 283
Chevy 302 & 327
Chevy 348 & 409
Chevy 350 & 400
Chevy 396 & 427
Chevy 454 thru 512
Chrysler Hemi
Chrysler 273, 318, 340 & 360
Chrysler 361, 383, 400, 413, 426 & 440
Ford 289, 302, Boss 302 & 351W
Ford 351C & Boss 351
Ford Big Block

B.B. RESTORATION SERIES

Auto Restoration Tips & Techniques
Basic Bodywork Tips & Techniques
BMW '02 Restoration Guide
Classic Camaro Restoration
Chevrolet High Performance Tips & Techniques
Chevy Engine Swapping Tips & Techniques
Chevy-GMC Pickup Repair
Chrysler Engine Swapping Tips & Techniques
Engine Swapping Tips & Techniques
Land Rover Restoration Tips & Techniques
MG 'T' Series Restoration Guide
MGA Restoration Guide
Mustang Restoration Tips & Techniques

MOTORCYCLING

B.B. ROAD TEST SERIES

AJS & Matchless Gold Portfolio 1945-1966
BMW Motorcycles Gold Portfolio 1950-1971
BMW Motorcycles Gold Portfolio 1971-1976
BSA Singles Gold Portfolio 1945-1963
BSA Singles Gold Portfolio 1964-1974
BSA Twins A7 & A10 Gold Portfolio 1946-1962
BSA Twins A50 & A65 Gold Portfolio 1962-1973
BSA & Triumph Triples Gold Portfolio 1968-1976
Ducati Gold Portfolio 1960-1973
Ducati Gold Portfolio 1974-1978
Ducati Gold Portfolio 1978-1982
Harley-Davidson Sportsters Pref. Port. 1965-1976
Harley-Davidson Super Glide Perf. Port. 1971-1981
Harley-Davidson FXR Series Perf. Port. 1982-1992
Honda CB750 Gold Portfolio 1969-1978
Honda CB500 & 550 Fours Perf. Port. 1971-1977
Honda CB350/400F Performance Portfolio 1972-78
Honda Gold Wing Gold Portfolio 1975-1995
Honda CBX 1000 Gold Portfolio 1978-1982
Kawasaki Z1 900 Performance Portfolio 1972-1977
Laverda Gold Portfolio 1967-1977
Moto Guzzi Gold Portfolio 1949-1973
Norton Commando Gold Portfolio 1968-1977
Suzuki GT750 Performance Portfolio 1971-1977
Triumph Bonneville Gold Portfolio 1959-1983
Vincent Gold Portfolio 1945-1980
Yamaha RD350/400 Performance Portfolio 1972-79

B.B. *CYCLE WORLD* SERIES

Cycle World on BMW 1974-1980
Cycle World on BMW 1981-1986
Cycle World on Ducati 1982-1991
Cycle World on Harley-Davidson 1962-1968
Cycle World on Harley-Davidson 1978-1983
Cycle World on Harley-Davidson 1983-1987
Cycle World on Harley-Davidson 1987-1990
Cycle World on Harley-Davidson 1990-1992
Cycle World on Honda 1962-1967
Cycle World on Honda 1968-1971
Cycle World on Honda 1971-1974
Cycle World on Husqvarna 1966-1976
Cycle World on Husqvarna 1977-1984
Cycle World on Kawasaki 1966-1971
Cycle World on Kawasaki Off-Road Bikes 1972-1979
Cycle World on Kawasaki Street Bikes 1972-1976
Cycle World on Norton 1962-1971
Cycle World on Suzuki 1962-1970
Cycle World on Suzuki Off-Road Bikes 1971-1976
Cycle World on Suzuki Street Bikes 1971-1976
Cycle World on Triumph 1967-1972
Cycle World on Yamaha 1962-1969
Cycle World on Yamaha Off-Road Bikes 1970-1974
Cycle World on Yamaha Street Bikes 1970-1974

MILITARY

B.B. MILITARY VEHICLES SERIES

Allied Military Vehicles No. 2 1941-1946
Complete WW2 Military Jeep Manual
Dodge Military Vehicles No. 1 1940-1945
Hail To The Jeep
Military & Civilian Amphibians 1940-1990
Off Road Jeeps: Civilian & Military 1944-1971
US Military Vehicles 1941-1945
US Army Military Vehicles WW2-TM9-2800
VW Kubelwagen Military Portfolio 1940-1990
WW2 Jeep Military Portfolio 1941-1945

20/13049

CONTENTS

ACKNOWLEDGEMENTS

Regular buyers of Brooklands Books will know that we have published several volumes dealing with the Camaros. One of the most popular was *Camaro Muscle Cars, 1966-1972*, which is now out of print. We thought it was better to put out this much enlarged Muscle Portfolio rather than merely to reprint the original book, particularly as interest in these cars has been on the increase recently and because we have uncovered a stack of material for which we simply did not have room in the earlier book. We should mention here that none of the stories in this book is duplicated in our recent *Chevrolet Camaro SS and Z-28, 1966-1973*, which remains in print.

As usual, we owe a debt of gratitude to the owners of the copyright to the material reproduced in this book. Our sincere thanks therefore go to the publishers of *Autocar, Car and Driver, Car Craft, Car Life, Cars, Hot Rod, Motorcade, Motor Sport, Motor Trend, Motor Trend Buyers' Guide, Popular Mechanics, Road & Track, Road Test and Sports Car Graphic*. We are also grateful to motoring writer James Taylor for his introductory paragraphs.

R.M. Clarke

The period which this book covers is still the most exciting one for Camaro enthusiasts, of it is the period of the big-block versions. After 1973, the Camaro was more image than real muscle car, and the fuel crises of the 1970s ensured that it would stay that way.

When the Camaro was introduced in the fall of 1966, it was an immediate hit, even though it was two years behind the Ford Mustang which had virtually established the market it was aimed at. Like the Mustang, the Camaro was available with an enormous range of options, so that buyers could equip it to suit their personal preferences. And inevitably, high-performance packages were among those options, for this was the great muscle car era.

As the stories in this book show, the Camaro quickly established itself as one of the great muscle cars. These stories tell of the big 427s, the Z-28s, the astonishing ZL-1, and of a few specially-prepared examples as well. In the days when there was no substitute for cubic inches, the name of Camaro was one to be reckoned with, and it still has a magical ring to it, twenty years after the last muscle Camaro was built.

For those who own a muscle Camaro, this book is essential. For those who dream about owning one, this book is addictive. And for those who like to have historical reference material on their bookshelves, this book is invaluable.

James Taylor

427 CAMARO

Corvette's Muscular Mover Provides
The Power for a Cobra-Style Transformation

BY JIM WRIGHT

THE CAMARO seems destined to be a second-thought car. For a time, Chevrolet preferred to ignore Ford's highly successful Mustang, but then had second thoughts. The Camaro was designed and built. The announcement of the Camaro stated the 350-cu. in. engine would be the top option. Meanwhile, with advent of the 1967 Mustang, Ford stated its top option would be the 390-cu. in. engine. On second thought, a month or so later, Chevrolet announced the 396-cu. in. engine option. Then, Carroll Shelby made a few announcements of his own, offering a package called the GT-500, his version of the Mustang powered by Ford's 428-cu. in. engine.

The Chevrolet factory ignored this one completely, but two of its dealers didn't. Nickey of Chicago and Dana of South Gate, Calif., had simultaneous second thoughts and started to build their own versions of the Camaro—powered by 425-bhp/427-cu. in. Chevrolet Corvette engines. In a way, they have trumped Shelby and Ford because the latter's 428-cu. in. engine is rated at only 355 bhp.

This action on the part of the two Chevrolet dealers wasn't surprising, considering their performance backgrounds. Nickey (with a mirror-image K) long has been active in stock, sports car and drag racing. Recently the firm announced a tie-in with Bill Thomas/racing cars of Anaheim,

Calif., for the purpose of building the 427 Camaros as well as merchandising a line of Chevrolet performance equipment and accessories.

Dana Chevrolet is a long-time operation that recently changed hands. One of the new partners, Peyton Cramer, for several years was associated with Shelby-American and was instrumental in initiation of the Mustang GT-350 project. Cramer understands hybrids. Rawhiding the Dana 427 Camaro project for Cramer is another Shelby works alumnus, Don McCain. His specialty is drag racing. McCain spent the past few years putting various versions of Cobra and Mustang GT-350 at the head of their respective drag racing classes.

CL wasn't able to have a go at the Nickey version, but test crewmen were able to thoroughly familiarize themselves with the Dana offering. And, some people think they have problems keeping the 350 Camaro at low altitude. They haven't seen anything. The 427 Camaro is just *too* much—in more ways than one.

The particular car tested was built by Dana for the Bardahl Oil Co., which plans to use it for show and as a high-speed test vehicle for numerous lubricant products. As such, the car carried a few more accessories than one usually would encounter on the average Dana 427 Camaro, mainly because of price. The base price of the

Dana package is $4495, but the extras raise the Bardahl car to $5500.

The Dana 427 price includes the base car, which is the 350 SS model. Standard 350 SS performance equipment includes Chevrolet's idea of heavy-duty suspension, increased capacity radiator (17 qt. rather than 13 qt.) and 14 x 6 wheels mounting D70-14 Firestone Wide Oval tires.

To this, as part of the $4495 package, Dana adds the 427/425 engine, Muncie 4-speed transmission, 3.55:1 Positraction rear end, metallic brakes, a set of headers with dual exhaust system, heavy-duty clutch and pressure plate with NHRA-approved scatter-shield, chrome-plated valve covers and air cleaner, 8500-rpm electronic tachometer, and dash-mounted oil pressure and water temperature gauges.

Extras on the Bardahl car, many of which are more for show than go, include custom interior, Rally Sport package, vinyl roof, custom steering wheel, tinted glass, push-button radio, appearance and lighting groups, front disc brakes, quick-response steering, Traction Masters and F70-14 Wide Tread tires. The last four items are 100% desirable choices from any enthusiast's standpoint.

DANA PEGS the unit package price at $1078, which makes allowance for the 350-cu. in. engine that is sold over the counter at $475, less starter

JUST AS IF THE engine compartment had been designed for it originally, the Corvette 427/425 fits snugly in place. Low-restriction paper element air cleaner and polished aluminum rocker covers are standard.

427 CAMARO

and alternator. This means the Camaro 350 SS, as Dana orders it, lists at $3417.

Nickey 427 Camaros carry a list price of $3711, which doesn't include special exhaust headers, heavy-duty clutch, pressure plate and approved scattershield, or extra instrumentation as does the higher base price of the Dana 427 Camaro.

The 427/425 engine used by both Dana and Nickey actually is a 1966 Corvette version of the 427 (RPO L72) that is ordered new from the factory. This is the single 4-barrel, mechanical lifter camshaft version. Compression ratio is 11:1. The engine achieves its rated 425 bhp at 5600 rpm and 460 lb.-ft. of torque at 4000 rpm. For the Camaro conversion, it appears a much better choice than any of the 1967 427s—390 bhp, single quad, hydraulic lifter camshaft; 400 bhp, three 2-barrel carburetors, hydraulic lifter camshaft; or 435 bhp, three 2-barrel carburetors, mechanical

lifter camshaft. It is better than the first two because it develops greater bhp and is better than the last—even with 10 fewer bhp—because a single 4-barrel is less trouble than three 2-barrel carburetors. It is also a slightly more responsive engine in the middle and upper ranges because of its longer cam timing—54-102, 102-54; 336° duration; 108° overlap; and 0.5197-in. lift compared with 44-92, 86-36; 316/302° duration; 80° overlap; and 0.5197-in. lift for the 1967 435-bhp engine. The 427/425 engine also has a broader torque curve, peaking out at 4000 rpm as compared with 3600 rpm for the 435-bhp engine.

Both engines share the basic internal parts—extruded aluminum pistons, forged steel crankshafts and drop-forged steel connecting rods. Intake and exhaust valves are identical for each engine—2.19-in., 1.72-in. diameter respectively. Valve spring pressures are 94-106 psi closed and 303-327 psi open, identical for either.

Swapping engines is a snap, because the Camaro engine compartment obviously was designed to accept the 396 and external dimensions of the 396 and 427 are exactly alike. The 427 engines arrive at the Dana shop minus carburetor, starter, alternator, pressure plate, clutch disc and bell housing. Starters and alternators are taken from 350-cu. in. engines. Other parts are added from stock. The big Holley carburetor is rated at 785 cfm and handles 427 engine demands. When the 427 is installed it is simply a matter of extracting the original engine and bolting in the replacement. There's plenty of room in the compartment for ease in maintenance and servicing.

THE RESULTING combination is not the best for 'round-town junketing. The fairly large Goodyear tires and absence of power-assisted steering make the 427 Camaro very difficult to park. If the driver doesn't have a set of well developed shoulder muscles, he will after a few sessions.

The throttle was approached with caution. The test car was fitted with a 3.55:1 rear axle, not an extreme ratio, but numerically high enough so full throttle in any gear produced a prodigious amount of wheelspin. Even if the car were eased out and not trod upon until well under way, as soon as the secondary throttles were opened and the engine climbed upon its camshaft at about 3500 rpm, the wheels broke loose. This occurred in every gear except fourth, which illustrates the tremendous amount of torque available from the Chevrolet 427/425 engine.

This is one engine that requires 2000 rpm be maintained for town-type driving. The camshaft is on the wild side and doesn't pull well in any gear below this engine speed. In traffic, the 30-mph city variety, fourth gear seldom was used.

During the acceleration runs at Carlsbad (Calif.) Raceway it was very difficult to get the Camaro off the line with any semblance of order. For some reason, not explained, the test car once had had a Positraction differential, but did not when received by *CL.* Such help was sorely needed. First gear, for all practical purposes,

was useless. Wheelspin was less violent in second gear, but still occurred throughout the range of the gear. Halfway through third gear, the tires finally bit solidly. Fourth gear produced only initial wheelspin. Traction Masters completely eliminated spring wrap-up, even in first gear where wheelspin was most pronounced. Very little effort was required to keep the car in a straight line.

Acceleration times were very good, in spite of the lack of traction, but really only indicate the 427 Camaro's potential. Chassis modification and suitable tires should easily knock a full 2 sec. off the car's 14.2 sec. e. t. in the quarter-mile. Zero to 60 mph probably would be under 5.5 sec. and the 0-80 mph time around 7.5 sec. Even without chassis tuning, the 427 Camaro will keep pace with just about anything operating on the streets of America today.

THE TEST car was equipped with optional front disc brakes and it was all they could do to meet the needs of the car. With standard linings, the

added weight of the 427 engine presents a few problems. The 23 ft./sec./ sec. stop from 80 mph was up to the usual standards test crewmen have come to expect from discs. The second panic stop indicated that the brakes were being overloaded and fade was apparent as the deceleration rate dropped to 21 ft./sec./sec. Some trouble was experienced with rear brakes locking. The brake system employs a pressure-proportioning valve between front and rear brakes to keep the rear from receiving enough pressure to cause locking. This works well in most cases, but with the Dana Camaro the added weight of the 427 caused an increase in the amount of apparent rear-to-front mass transfer during hard braking and removed more mass from the rear tires than anticipated. The result is that rear brake lock occurs before it should, thereby keeping the rate of deceleration from exceeding 23 ft./sec./sec. Here, again, a certain amount of chassis tuning might be able to alleviate the problem. Fade probably could be cured by switching to competition-

HEADERS nestle between engine and fender panels, help produce added power.

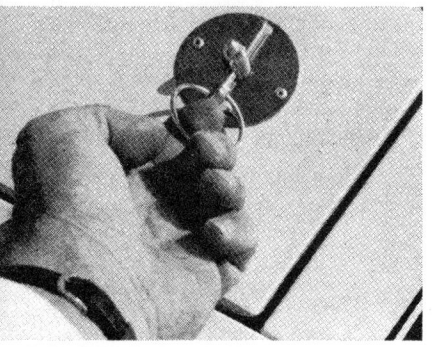

PIN LOCKS for hood hold-downs are a racing styled dress-up accessory.

JIM WRIGHT PHOTOS

ELECTRIC TACHOMETER atop steering column and oil pressure and coolant temperature gauges are standard additions for 427 Camaro.

type brake pads at the front. The rear drums already are equipped with metallic linings as part of the Dana package. In any event, the front brakes did not lock and all the test stops were accomplished in fairly straight lines with the barest minimum of steering correction required.

The test Camaro was equipped with quick steering option, which added to the car's overall responsiveness. One serious drawback to the 427 Camaro is a considerable degree of understeer, much greater than displayed by the factory Camaro. It appears that engine weight is the culprit once again. The chassis is designed for the weight of the 350-cu. in. engine and front and rear suspension geometry is set accordingly. Additional weight upsets the original balance. A larger front antiroll bar would help. The one supplied has a diameter of 0.6875, which is identical to the one used on a Camaro with a lighter 327 engine. The addition of a rear antiroll bar, as in Oldsmobile's 4-4-2, also might help matters by increasing roll stiffness at the rear and less forward mass transfer during cornering. The test car also exhibited very little jounce/rebound control, though so-called heavy-duty shock absorbers were factory installed. They were especially poor on rebound. If slight dips or bumps were encountered on high speed freeway curves, lack of rebound control, coupled with the high degree of understeer, never failed to produce a few anxious moments for the driver. Test crewmen declined to wring out the suspension on mountain roads, discretion being the better of valor.

The supreme test for a car of this type came when it was shown off to a certain well-known engine builder. His shop had the usual contingent of drag racers hanging about as the 427 Camaro drove up. Because all involved are prominent in the sport and supposed to be old enough to know better, names

1967 CHEVROLET
DANA 427 CAMARO

DIMENSIONS

Wheelbase, in.	108.1
Track, f/r, in.	59.0/58.9
Overall length, in.	184.6
width	72.5
height	51.0
Front seat hip room, in.	2 x 20.5
shoulder room	56.7
head room	37.0
pedal-seatback, max.	40.5
Rear seat hip room, in.	54.8
shoulder room	53.8
leg room	30.5
head room	36.7
Door opening width, in.	41.2
Floor to ground height, in.	10.0
Ground clearance, in.	6.3

PRICES

List, FOB factory............$4495
Equipped as tested............5500
Options included: RS package, vinyl roof, custom interior, tinted glass, radio, appearance and light groups, power disc brakes, Traction Master torque arms.

CAPACITIES

No. of passengers	5
Luggage space, cu. ft.	8.3
Fuel tank, gal.	18.5
Crankcase, qt.	5.0
Transmission/diff., pt.	3.0/4.0
Radiator coolant, qt.	22.0

CHASSIS/SUSPENSION

Frame: Unitized body; front subframe.
Front suspension type: Independent by s.l.a., coil springs, telescopic shock absorbers, ball-joint steering.
ride rate at wheel, lb./in......125
antiroll bar dia., in......0.6875
Rear suspension type: Live axle, Hotchkiss drive; single-leaf parallel springs, telescopic shock absorbers, torque arms.
ride rate at wheel, lb./in......131
Steering system: Semi-reversible recirculating ball nut gear, parallelogram linkage.
gear ratio......24.0
overall ratio......21.6
turns, lock to lock......3.5
turning circle, ft. curb-curb....37.0
Curb weight, lb......3368
Test weight......3728
Weight distribution, % f/r...59/41

BRAKES

Type: 2-circuit hydraulic with tandem master cylinders; caliper discs, front; duo-servo shoes in composite drums, rear.
Front rotor, dia. in......11.0
Rear drum, dia. x width....9.5 x 2.25
total swept area, sq. in......332.4
Power assist: Integral, vacuum
line psi @ 100 lb. pedal......n.s.

WHEELS/TIRES

Wheel size......14 x 6JK
optional size available......14 x 5J
bolt no./circle dia., in......5/4.75
Tires: Goodyear Wide Tread
size......F70-14
recommended inflation, psi......26
capacity rating, total lb......4840

ENGINE

Type, no. cyl......ohv, 90° V-8
Bore x stroke, in......4.25 x 3.76
Displacement, cu. in......426.506
Compression ratio......11.0
Rated bhp @ rpm.....425 @ 5600
equivalent mph......120
Rated torque @ rpm......460 @ 4000
equivalent mph......86
Carburetion......Holley, 1x4
barrel dia., pri./sec......1.686/1.686
Valve operation: Mechanical lifters, pushrods, overhead rockers.
valve dia., int./exh......2.19/1.72
lift, int./exh......0.5197/0.5197
timing, deg......54-102, 102-54
duration, int./exh......336/336
opening overlap......108
Exhaust system: Headers, dual mufflers.
pipe dia., exh./tail......2.50/2.00
Lubrication pump type......gear
normal press. @ rpm.50-75 @ 2000
Electrical supply......alternator
ampere rating......37 @ 12 V.
Battery, plates/amp. rating....66/61

DRIVE TRAIN

Clutch type: Diaphragm; centrifugal disc.
dia., in......11.0
Transmission type: Manual 4-speed.
Gear ratio 4th (1.00) overall......3.55
3rd (1.27)......4.52
2nd (1.64)......5.83
1st (2.20)......7.82
synchronous meshing?......4 forward
Shift lever location......floor console
Differential type: Hypoid; overhung pinion.
axle ratio......3.55

won't be recorded. After one and all had looked over the car, the decision was that nothing would do but all should pile into the car and head for a nearby drive-in restaurant, the purpose of this impromptu excursion being to "blow the kids' minds."

As it was evening, the drive-in lot was was full of the usual crowd. Cars and people of all description were gathered. Tight little groups stood about discussing whatever it is tight little groups discuss. Pulling into the lot, we selected first gear and the Camaro was allowed to idle on through. The radical camshaft produces a nasty lope at first gear idle—very effective. From time to time as the car pulled abreast of first one group, then the other, its occupants shouted, "Go for pinks," or other such archaic phrases. These pointed challenges were met with rather slack-jawed silence. We cruised through twice more with like results. The kids' minds were blown! The nasty lope of the engine could have done it, though several of those involved suspect it may have been the sight of a carload of older types trying to regain the automotive pleasures of their lost youth.

In any event, the 427 Camaro cruised off into the night, unchal-lenged, unmolested and, probably, un-ticketed.

One of the participants on the way back to the shop said, "There we were —all of us—17 years old again." And that's really what a car like this is all about. Of course, if one happens to really be 17, he probably won't notice anything other than the fact that the Dana Camaro is an exceptionally fast car.

It's quite a feeling to be 17 again, but something a little less spectacular, with a little more emphasis on *overall* performance and a little less on *raw, brute horsepower* is *CAR LIFE*'s preference. But then again, with a little help in the suspension department. . .■

CAR LIFE ROAD TEST

ACCELERATION & COASTING

MPH — **ELAPSED TIME IN SECONDS**
(graph: 4th, SS ¼, 3rd, 2nd, 1st; MPH axis 10–120; time axis 5–45)

CALCULATED DATA

Lb./bhp (test weight)	8.78
Cu. ft./ton mile	187
Mph/1000 rpm (high gear)	21.4
Engine revs/mile (60 mph)	2800
Piston travel, ft./mile	1750
Car Life wear index	49.2
Frontal area, sq. ft.	20.5
Box volume, cu. ft.	395

SPEEDOMETER ERROR

30 mph, actual	29.0
40 mph	38.0
50 mph	47.0
60 mph	56.0
70 mph	64.0
80 mph	74.0
90 mph	85.0

MAINTENANCE INTERVALS

Oil change, engine, miles	6000
trans./dif.	as req.
Oil filter change	6000
Air cleaner service, mo.	6
Chassis lubrication	6000
Wheelbearing re-packing	as req.
Universal joint service	none
Coolant change, mo.	24

TUNE-UP DATA

Spark plugs	AC 43N
gap, in.	0.033-0.038
Spark setting, deg./idle rpm.	8/800
cent. max. adv., deg./rpm.	28/4600
vac. max. adv., deg./in. Hg.	15/12
Breaker gap, in.	} magnetic
cam dwell angle.	} pulse
arm tension, oz.	}
Tappet clnc., int./exh.	0.024/0.028
Fuel pump pressure, psi	5.7-7.0
Radiator cap relief press., psi	15

PERFORMANCE

Top speed (6000), mph	130

Shifts (rpm) @ mph, manual

3rd to 4th (6000)	101
2nd to 3rd (6000)	78
1st to 2nd (6000)	58

ACCELERATION

0-40 mph, sec.	3.7
0-50 mph	4.9
0-60 mph	6.3
0-70 mph	7.7
0-80 mph	8.4
0-90 mph	11.4
0-100 mph	14.0
0-110 mph	17.3
Standing ¼-mile, sec.	14.2
speed at end, mph	102
Passing, 30-70 mph, sec.	5.4

BRAKING

(Maximum deceleration rate achieved from 80 mph)

1st stop, ft./sec./sec.	23
fade evident?	slight
2nd stop, ft./sec./sec.	21
fade evident?	yes

FUEL CONSUMPTION

Test conditions, mpg	n.a.
Normal cond., mpg	n.a.
Cruising range, miles	n.a.

GRADABILITY

4th, % grade @ mph	20 @ 90
3rd	30 @ 74
2nd	40 @ 56
1st	off scale

DRAG FACTOR

Total drag @ 60 mph, lb.	120

CAMARO

'67's MOST **ANTICIPATED** CAR

Chevrolet's biggest gun for 1967 is, of course, the long-awaited Camaro — a tag unexpectedly bestowed. But if the title is unexciting, the car is just the opposite. Specifically introduced to quell Ford's Mustang sales, the Camaro embodies styling that has come to exemplify "sportiness" and which Chevrolet claims was founded in the U.S. by Corvette; long hood, short deck, to put it briefly.

For openers, the Camaro is available as either a convertible or a 2-door hardtop, each in three basic guises—standard, with a Rally Sport Exterior Package, or with a Super Sport 350 Package. With five engines on tap (from a 230-inch in-line 6 to an all-new 350-inch TurboFire V-8), three transmissions (3- or 4-speed sticks and PowerGlide) and a host of other interior and exterior options, the Camaro can be virtually tailor-made to an owner's individual specifications.

Chassis layout is conventional; i.e., front-mounted engine, rear wheel drive. The Camaro's maximum four passengers ride on a 108-inch wheelbase. The ab-

breviated 36.6-inch frontal overhang and fairly long 40.1-inch rear overhang stretch the Camaro to a total length of 184.7 inches, 29 inches shy of the full-sized Chevy, thirteen less than the Chevelle, but an inch and a half longer than the Chevy II — some of whose chassis components the Camaro shares.

Indications stemming from a pre-introductory drivetest give promise of a very road-worthy automobile that should give the horse a good run for the money.

Camaro coupe is standard version of new Chevrolet sportster. Front end sheetmetal utilizes bolted-on, rubber isolated construction method. Optional SS package includes special headlight mounting behind grille doors which are electrically opened to reveal lamps when needed.

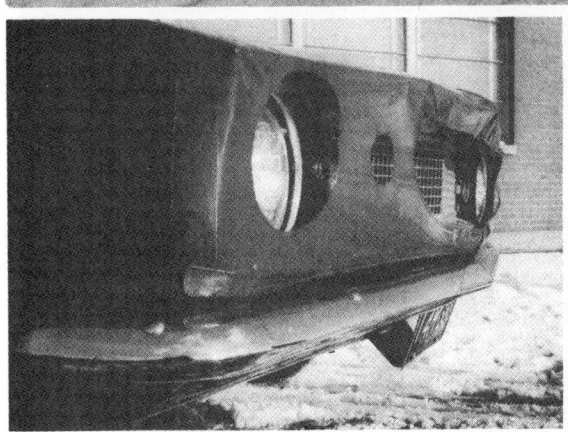

Detroit sometimes has more secrecy than the CIA. The Camaro project was no exception. Rumors have prevailed for months that Chevy had their fingers in the Mustang bag. And, indeed, they were working on an exciting automobile for '67. They tagged it the "F" car during development while most auto magazines dubbed the new vehicle "Panther." Chevy, sticking with the letter C, finally named it the Camaro. Above are some views of the Camaro in its early testing stages. Chev Engineering Test and Development camouflaged the front end to foil industrial spies with telephoto-lensed cameras. At top of page is pre-production test car.

Camaro Super Sport coupe with padded top illustrates long hood—short deck designing prevalent on several '67 cars. Wheelbase is 108 in. with length measuring 184.7 in.

Styling of an automobile — especially a completely new car — is a long process involving much effort before the car finally takes shape. Above is a craftsman forming in clay what stylists had put on paper during development.

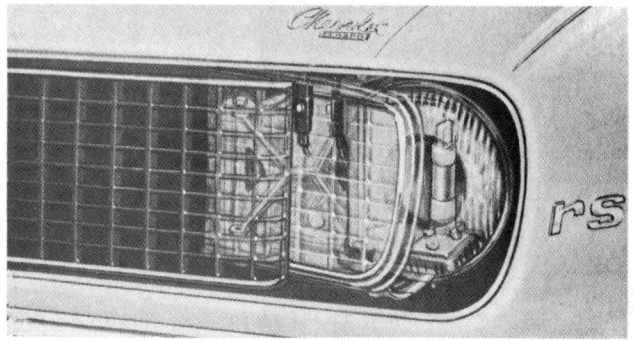

The optional Rally Sport Exterior package includes headlights concealed behind grille. When actuated, grille pieces forming "doors" in front of lamps are electrically moved to the side, allowing normal headlight usage.

Standard Camaro convertible borrows much, in this view, from Corvair styling. If desired, buyer can build his Camaro as fine personal car or as ultra high performance SS screamer.

Chevy's 350 cubic inch V8 engine is available exclusively in the Camaro. The Super Sport package includes special hood treatment and paint striping which they call "bumble-bee" paint along with handling equipment for chassis and 350 engine.

Camaro interior is attractively designed and adds sporty flavor to car. Full line of options is offered for interior.

Optional Deluxe Interior package features molded vinyl panel with integral armrests and door handles mounted in depression.

TWO BODY STYLES, AN ENGINE/ TRANSMISSION CHOICE TO FILL EVERY NEED, AND BUCKETS OF OPTIONS ADD UP TO A FUN CAR

Camaro

BY CHEVROLET

BY STEVEN KELLY

M/T ROAD TEST

Car enthusiasts awaited the introduction of the Camaro with the anticipation of a little boy at Christmas who had a new bicycle on order from Santa Claus. Like the little boy, no one was sure what it was going to look like, but they were rather sure they'd like it. □ The expected 300,000 '67 Camaro buyers have been given a long list of options to choose from, and a base price low enough to encourage just about everyone to at least take a look at the car. □ Chevrolet General Manager E. M. "Pete" Estes has described the Camaro as a "4-passenger package of excitement." A look at the available power teams and comfort and/or performance options is cause enough to re-phrase this statement to "4-passenger package of choice." □ This, of course, is by design — not accident. One of Mustang's success secrets was that the same car could be equipped and priced for the low-budget buyer or the buyer who usually ordered everything possible. and didn't mind paying. This brought a lot of high-price buyers over to the Mustang, and Chevrolet hopes it'll work for them, too. □ Human nature sells a lot of cars, and the young-at-heart emphasis placed on the Camaro is going to get a lot of attention from oldsters who refuse to let their minds age, as well as the youngsters under 25 who spend most of the automotive dollar. □ Chevrolet's engineering and styling departments worked together in prescribing the basic form and dimensions for the Camaro. It was then up to the stylists to create a design which was both functional and pleasant to the eye.

14

Coupe and convertible bodies are the only style choices for Camaro. Either style can be fitted with SS-350 V-8, recognizable by "bumblebee" stripe around front end. Rally Sport grille is shown here on coupe, while convertible has standard item. Various trim offerings include three wheel cover designs.

Front seatback latches prevent them from flopping forward during quick stops, and can be easily reached for entry or exit. Safety wire stops seatback before it has a chance to strike dashboard top. Optional center console contains rear compartment light—operated by door switches—and ash tray.

camaro

Standard-equipment grille with closely placed head- and parking lights can be replaced with optional Rally-Sport package. RS grille has hide-away lights concealed by electrically operated doors that pivot inward, while lights remain stationary. Parking lights are then relocated in the lower valance panel.

Aerodynamics became a prime consideration of the stylists. A car as small and light as the Camaro can be highly affected by wind gusts if not properly designed. In their search for an aerodynamically sound style, Chevrolet enlisted the aid of a computer.

The computer didn't actually design the car, but it quickly affirmed or denied the progressing work. This made it very easy to instantly check the work before going on to another portion of the car.

After the human and mechanical brains had completed their design work, a scale-model mock-up was built and taken to a Dallas, Texas, wind tunnel. Computers were present during this operation also, receiving information gained in the wind tunnel and transferring it into a workable answer. The final results showed that only a minor alteration to the front valance (sheetmetal below the bumper) would give a smooth and aerodynamic shape to the body.

All this work could have been accomplished manually, but it would have taken hundreds of times longer and the resulting answer may not have been as accurate.

body and chassis design reflects sports car image

Two body styles are offered on the Camaro: a coupe and a convertible. The convertible's unit-body construction is heavier, due to reinforcements in the floor pan and the side panels. It also gains a hundred pounds of weight from the addition of vibration dampeners placed in each corner of the car.

All Camaros have unit-body construction, using a separate forward frame which attaches to the body just below the front-seat area and extends forward to the nose. The engine and transmission are hooked to the sub-frame.

How's this for choice? Depending on the transmission, you can also choose the shifter location. Standard 3-speed manuals are controlled from the column, while heavy-duty 3-speeds and 4-speeds have floor-mounted sticks. An optional console can be wrapped around either of the manual shifters. Powerglides have a column-mounted selector and can be ordered floor-mounted.

The Rally-Sport option accomplishes rear-end changes as well as frontal. Center-fill gas cap is identified with "RS" letters, and tail lights are solid red with flat black trim. Standard tail lights have bright trim with back-up lights integrated. Back-up lamps on RS models are positioned in lower valance.

To attain the desired ride characteristics, ride and handling engineers also enlisted the services of a computer. Information received from the interviews with private and professional drivers, regarding their ride preferences for a "sports" car, was fed into the programmed computer. A few clicks, buzzes and flashing lights — and the computer arrived at a final answer on how the Camaro should ride.

Not only did the computer find out how it should handle, it told how to make it handle the way it should. Thousands of chassis combinations exist, and physically changing them all would take longer than the total working years of the average Chevrolet engineer. The final suspension combination was within a few degrees of being 100% correct. Here again, the work could have been accomplished manually, but it's doubtful if it would have turned out as accurate.

Front suspension of the Camaro is independent, with coil springs. The engine sits almost directly over the centerline of the front end, giving excellent weight distribution and near-neutral handling. The rear suspension consists of a conventional one-piece axle housing, sprung with single-leaf springs as in the Chevy II. There are several different axle ratios available for the smaller-engined cars, and seven different ones for the SS-350. The lowest is a 4.88-to-1, designed expressly for the man who wants to get away quick, and the highest, 3.07-to-1, favors cruising economy.

All SS-350 models and 327 4-barrel-equipped cars have an extra traction bar attached to the rear end to eliminate excessive wheel hop found on early test cars.

engines range from subtle 6s to all-out 8s

Standard engine in all 6-cylinder models is a 230-cubic-inch L-6, rated at 140 hp. For a few dollars extra, buyers may order the bigger, optional 250-cubic-inch in-line 6, rated at 155 hp. Both engines will run smoothly on regular gas, due to the low, identical, 8.5-to-1 compression ratio.

A 210-hp, 327-cubic-inch V-8 is standard in V-8 Camaros. A 2-barrel carb and 8.75-to-1 compression ratio are featured, making it possible to run on regular gas. A 275-hp version of the 327-cubic-inch engine is offered, and it has a single 4-barrel carb and 10-to-1 compression.

A Camaro exclusive is the 350-cubic-inch V-8 which comes as part of the SS-350 option. This engine is supplied with a 4-barrel carb and is modestly rated at 295 hp. Torque is listed at 380 pounds-feet at 3200 rpm. The stroke is the same as the 327 V-8's, but bore is increased to 3.48, vs. the 3.25 of the smaller engine.

All engines are available with either a 4-speed manual transmission or a Powerglide automatic. Three-speed, all-synchro manual gearboxes are standard, and a 3-speed, heavy-duty box can be ordered with the 350-inch engine choice only.

Camaro

SS-350 on left dropped quick e.t.'s and fast ¼-mile speeds with ease. Extremely smooth 327 V-8 goes slower — but cheaper.

Our test cars were both coupes, each had disc brakes, deluxe interiors, special instrumentation and the Rally-Sport option. The Rally-Sport package consists of "RS" emblems, all-red tail lights with back-up lights in the valance, and a grille with concealed headlights.

The SS-350 option includes heavier suspension components, Wide-Oval tires and of course, special "SS" emblems, a "bumblebee" stripe up front, and simulated hood scoops.

Our SS-350 test car really surprised us. Quarter-mile times were exceptionally good, especially considering the over-3500-pound weight, with two passengers plus test equipment aboard. We've tested some comparably equipped cars that were considerably slower.

Some finesse with the gas pedal was needed to get the SS-350 off to a good start. High-rpm runs produced excessive wheelspin — though not much smoke from the Wide-Ovals — and slow times. Coming out with the tach needle resting just below 3000 rpm, and then stabbing it, produced the best times in the acceleration runs.

The shift linkage on the 4-speed Muncie box was firm, smooth and precise, meaning that when it was in first gear, it felt like it. There was no gear clashing, and full power shifts were possible.

The Powerglide-equipped car with its 210-hp V-8 proved to be as comfortable a car as the SS-350, and had the advantage of asking less of the driver. We made acceleration runs with the stick in the "D" position. It shifted from low to high at 4900 rpm — exactly — every time. The recommended red-line for this engine is 5000 rpm, and we weren't about to try shaving it any closer than the transmission could do on its own.

The disc brakes on both cars brought them to quick, safe stops, though it takes some "light-footedness" to keep from locking them up. Once we got used to light, slow pressure on the pedal, we found ourselves being gently slowed by the front disc/rear drum combination.

Because we worked long before public introduction date, we weren't able to take the cars off the test track at GM, and this kept us from doing any long-range mileage checks. We'd estimate the SS-350 to be capable of 12-13-mpg average, and at least 15 mpg for the 210-hp 327 V-8.

The combination of electric windows and the deluxe interior option eliminates all protrusions from the door panels and with it, the possibility of catching your clothes on a handle. The windows, of course, have a simple switch, while the door handles are neatly tucked away in "pockets" as a part of the option. Also included are rear-seat arm rests with ashtrays.

The Camaro is one of the most pleasurable cars of its size — or any other size — we've driven. Wide-ranging seat adjustment enables tall or short drivers to navigate with equal ease, and the tilting wheels adds driving comfort.

Enthusiasm for the Camaro comes easy. It invades the luxury kingdom of Cougar by offering more comfort options, and the sporty area of the Mustang by having at least the same amount — if not more — of enthusiast-oriented accessories. Even the Barracuda doesn't get away unscathed — Camaro convertibles and coupes can be equipped with a fold-down rear seat. /MT

M/T Road Test

car at a glance . . .

Sub-frame, unit-body construction . . . computer-verified styling and suspension . . . optional front disc brakes . . . two types of grilles, one with disappearing headlights.

how the car performed . . .

ACCELERATION (2 aboard)

	327	SS-350
0-30 mph	4.0 secs.	2.6
0-45 mph	6.8 secs.	5.1
0-60 mph	10.7 secs.	8.0
0-75 mph	16.7 secs.	11.4

TIME & DISTANCE TO ATTAIN PASSING SPEEDS:
40-60 mph:
  5.2 secs., 380 ft.   3.6 secs., 263 ft.
50-70 mph:
  6.6 secs., 580 ft.   4.1 secs., 360 ft.

STANDING-START QUARTER-MILE:
18.2 secs. and 77 mph   15.4 secs. and 90 mph

BEST SPEEDS IN GEARS @ SHIFT POINTS:
327
1st . . . . 60 mph @ 4800 rpm
2nd . . . . 72 mph @ 3000 rpm
(not maximum)

SS-350
1st . . . . 44 mph @ 5000 rpm
2nd . . . . 58 mph @ 5000 rpm
3rd . . . . 75 mph @ 5000 rpm
4th . . . . 56 mph @ 2500 rpm
(not maximum)

MPH PER 1000 RPM: (327) 24; (SS-350) 22.4

STOPPING DISTANCES:
(327) from 30 mph, 36 ft.; from 60 mph, 151 ft.
(SS-350) from 30 mph, 36 ft.; from 60 mph, 156 ft.

SPEEDOMETER ACCURACY

Calibrated Speedometer	30	45	50	60	70	80
Car's speedometer (327)	30	46	52	61	74	NA
(SS-350)	30	45	50	60	70	80

specifications . . .

	327	SS-350
ENGINE:	Ohv V-8	Ohv V-8
Bore & stroke (ins.):	4.00 x 3.25	4.00 x 3.48
Displacement (cu. ins.):	327	350
Horsepower:	210 @ 4600 rpm	295 @ 4800 rpm
Max. torque (lbs.-ft.):	320 @ 2400 rpm	380 @ 3200 rpm
Compression ratio:	8.75:1	10.25:1
Carburetion:	1 2-bbl.	1 4-bbl.
TRANSMISSION:	Powerglide 2-speed automatic	4-speed manual; all-synchro
FINAL DRIVE RATIO:	2.73:1	3.31:1

SUSPENSION: Front: Independent front with coil springs and concentric shock absorber. Link-type stabilizer. Rear: Salisbury-type rear axle housing with 2 single-leaf springs. Tubular shock absorbers.
STEERING: (327) Optional power — co-axial with semi-reversible, recirculating ball nut. (SS-350) Manual — semi-reversible, recirculating ball nut.
Turning diameter: 37.0 ft. curb to curb.
Turns lock to lock: 3.0 power assist; 4.0 manual.
WHEELS: Steel, short-spoke spider. 14-in. dia.
TIRES: (327) 7.35 x 14 rayon tubeless.
(SS-350) Wide-Oval nylon tubeless.
BRAKES: Dual-system hydraulic, duo-servo. Optional front discs or metallic linings.
Front disc, 11.0 in. dia. (both cars).
Rear drum, 9.5 in. dia. x 2.00 in. wide
FUEL CAPACITY: 18.5 gals.
BODY AND FRAME: Combination body-frame integral with separate forward portion ladder frame.
WHEELBASE: 108.1 ins.
TRACK: front, 59.0 ins.; rear, 58.9 ins.
OVERALL: length, 184.6 ins.; width, 72.5 ins.; height, 51.0 ins.
USABLE TRUNK CAPACITY: 8.3 cu. ft., coupes; 5.6 cu. ft., convertibles.
CURB WEIGHT: (SS-350) 3141 lbs.; (327) 3228 lbs.

MANUFACTURER'S SUGGESTED LIST PRICE: excludes state and local taxes, license, options, accessories, and transportation.

	Camaro V-8 (327 cu. ins., 210 hp)	Camaro 6 (230 cu. ins., 140 hp)
Sport coupe	$2572	$2466
Convertible	2809	2704

OPTIONS AND ACCESSORIES

Rally-Sport package	$105.35
Super Sport 350 package	210.65
Custom interior	94.80
4-speed transmission (V-8 models)	184.35
4-speed transmission (6-cyl. models)	115.90
Powerglide (V-8 models)	194.85
Powerglide (6-cyl. models)	184.35
Air conditioning	356.00
Front disc brakes	79.00
Special-purpose suspension (std. on SS-350)	10.55

IT WAS INEVITABLE! The question was who would be first to drop a 427 into a Camaro. With General Motors' unwritten policy limiting Chevrolet to ten horsepower per each 100 pounds on vehicles other than sports cars, the stock class Chevy enthusiast was struck a serious blow. It meant that a model such as a 3,000 pound Camaro would be limited to only 295 horsepower. With Ford and Chrysler putting their high horsepower engines in their small cars, Chevrolet fans might well be left pretty much "out to lunch."

There is one Chevrolet sales and racing group, however, that likes to field very competitive G.M. vehicles. Chicago's Nickey Chevrolet recognizes that sales increase when factory cars are seen running in open competition. Nickey has long been associated with automobile racing, having sponsored a variety of Corvette and Chev-powered sports cars for the past ten years and currently sponsors a pair of Chevy Vinegaroons (a much modified MK 10B Genie sports-racing cars) owned by "Bonanza's" Dan Blocker. Nickey sponsorship also backs match racer Dickie Harrell and the Colson & Wood gas dragster on the nation's drag strips.

With interest in drag racing reaching mammoth proportions, Nickey's Chevrolet recognized the potential market involved and added a complete array of high performance drag racing merchandise to their 20,000 square foot parts area. This addition brought their total parts to more than 140,000 items in stock, ready for any customer whether he was interested in racing or just keeping his stocker running as efficiently as possible.

Nickey's parts shelves stock high performance Chevrolet parts that some G.M. dealers don't even know the factory builds. It didn't take long, however, for the performance enthusiasts who had frequently received the cold shoulder from Chevrolet parts counters around the country to learn that Nickey's was the place to get Chevy speed goodies, whether it was factory high performance options or special components made by the speed equipment industry. With 27 people working in the parts department, it was easy to see how Nickey could rate as the largest Chevrolet parts dealer in the country.

The first Camaro had hardly slipped off the transport at the Nickey lot before Dick Harrell, the AHRA professional stock car points champion from Carlsbad, N.M., and now Performance Advisor at Nickey's, and parts manager Ken White got busy measuring the engine compartment. The new 350 cubic inch,

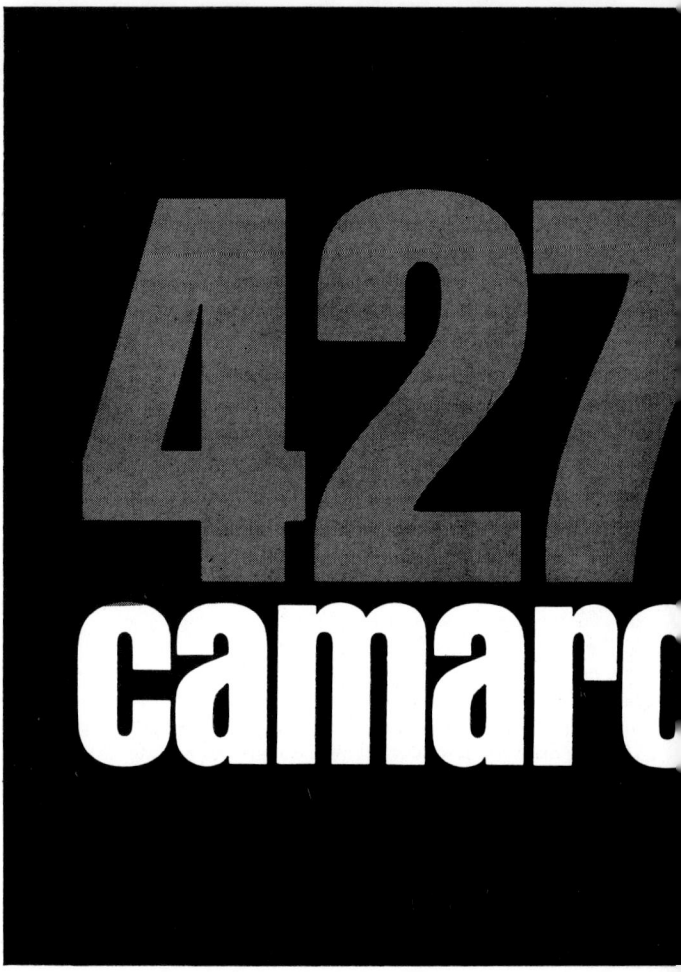

Nickey Chevrolet and Bill Thomas Race Cars combine to produce the wildest Camaro ever b dropping Chevrolet's biggest engine int the new GM featherweigh

LEFT ABOVE — Until Nickey made their 427 Camaro available, the 350 cubic incher was Camaro's biggest engine option. Based on the 283 block, the 350 left a lot of space for changes.

LEFT — All ready for a customer, the 427 fits neatly in place of the 350 engine. Simple change jumped horsepower from 295 to 435, adding only 90 pounds. Snorkle was chopped for clearance.

ABOVE LEFT — Mike Terrafino inst the big Holly on newly arrived 427 wh "old" 350 awaits return to parts warehou No internal carb changes were ma

ABOVE — Nickey Performance Advis Dick Harrell, watches Lou Anzelmo a Terrafino as they shift the 427 into pla Harrell coordinates conversion progra

BY DICK SCRITCHFIELD

295 horsepower "SS" engine was nice but it still wasn't the largest powerplant in the Chevy warehouse, so the crew at Nickey's was looking for something more.

To the racers at Nickey's, the Camaro was a natural. It was the same general size as the Corvair with just the right location of axles and engine to make it a hit as a quarter-mile sprinter. The combination was there, so why not drop in the biggest 427 cubic inch mill, one that would put out 435 horsepower in its factory trim? Here could be a car equally at home on the street or strip.

If the 350 engine could be swapped for the 427 with a minimum of problem by using stock parts, Harrell felt that this would be the answer to what the Chevy hot rodders wanted. They might even cause a few desertions from other camps.

Harrell found that by removing the hood, battery, radiator and 350 engine, the 427 assembly slipped right in, even to the exact matching of the engine mounts. No cutting, no bending, no welding in the engine compartment. Pretty slick!

Finding that the 427 would fit the engine compartment, Harrell got busy determining the type of car they could build for a customer and possibly keep it eligible to compete in a drag racing stock car class. Although the National Hot Rod Association doesn't classify anything built outside the basic factory as being stock, the American Hot Rod Association decided that there should be a place in its new Super Stock class for a Chevrolet product. With both Ford and MoPar planning hot engines for their sports-personal cars, General Motors would be clear out of the picture with their challenger limited to under 400 cubic inches. In view of Nickey's planned production, it was determined that Nickey-built 427 Camaros would be eligible for AHRA Super Stock competition. At this writing, these are the only 427 Camaros that will be allowed to run as a stocker.

To meet AHRA's new Super Stock class requirements, the car must have over 400 cubic inches, be stock bore and stroke, weigh a minimum of 3250 pounds, have an unaltered wheel-base, no engine relocation, run pump gasoline and have no lightweight components other than the hood. It sounds like a good class and should create a lot of excitement.

Looking over the available options and combinations offered by Chevrolet for the Camaro line, Harrell and White discovered that the SS 350 Camaro with heavy duty suspension and radiator, metallic brakes, four-speed Muncie transmission, positraction and red stripe tires would be the most economical beginning for a high performance combination street/drag machine. By making the engine swap before the customer took delivery, Nickey would be able to provide the customer with warranties as well as giving them the benefit of using the discarded 350 engine and other non-used items to help reduce parts and installation charges.

The heavy duty suspension includes stiffer front and rear springs, plus a front stabilizer to eliminate high speed cornering sway. With the larger, hard-running 427, Harrell felt the cooling capabilities should be increased so the optional air conditioning radiator, with its extra row of cooling tubes, was installed. The wide ratio Muncie four-speed provides ratios of 3.00:1 in first, 2.20:1 in second, 1.47:1 in third and 1:1 in fourth. For only $10 more you can have ratios of 2.54:1, 1.80:1, 1.44:1 and 1:1; first through fourth. And for those who prefer the automatic, the standard or super-duty Turbo-Hydramatics are available.

Posi-traction axle ratios are available in 3.31, 3.70, 4.10 and 4.56:1, giving the customer a complete variety of gears to cover his requirements for street, the drags, or both. Provided the Camaro is in stock with the ratio desired by the customer, there is no additional charge for any gear combination. If the ring and pinion must be changed, only a labor charge is added.

Chevrolet offers a pair of "427" blocks. The 427 cubic inch, 385 horsepower "economy" engine which comes with hydraulic lifters, 10.25:1 compression ratio and a cast iron intake manifold, and the 425 horsepower mill. Wanting to provide the

(continued on following page)

The 427 exhaust manifold and 350 head pipe have nothing in common, meaning the head pipe must go. Clearance around bottom of engine leaves lots of space for steering linkage to move.

ultimate in performance and reliability, Harrell chose the generally beefier construction of the 425 horsepower version, which features extra strong four bolt mains and full 360 degree oiling of the rod journals.

Standard with the engine is a solid lifter camshaft with a .517 lift, but a high performance version with hydraulic cam and lifters, providing a quieter, smoother running engine, is available for street use. Other good stuff inside includes forged steel rods, 11:1 compression heads and an aluminum intake manifold which Harrell outfits with a four-barrel Holley (No. 3886091) in place of the factory installed Rochester. By replacing the single four-barrel manifold and Holly with a trio of two-barrels on a hi-riser manifold, an extra ten horsepower can be gained.

To go a step further, a boost to 450 horsepower is obtained with an exclusive Nickey manifold holding two high performance Carter AFB four barrel carbs. Only problem here is that the stock hood won't close. But don't worry, Harrell and White have an answer for that. They have a competition scooped (ala 'Vette) fiberglass hood that fits over the multiple carburetor installations. If you're interested in saving weight, but want the hood to appear stock, Nickey can also provide you with a lightweight 'glass one for the regular carburetion set-ups.

The length of the "427" was the only major difference when compared to the "350." It extends two inches toward the front, ahead of the leading engine mounts.

By attaching the fan directly to the water pump pulley instead of using the spacer of the "350" engine, the fan mounts at the proper location in the shroud. An extra 90 pounds is added over the front suspension by the 427. To decrease weight, plus adding to the overall efficiency of the engine, the stock cast iron exhaust manifold can be replaced with Bill Thomas designed and built headers. When installed at the time of the swap, there is no additional labor charge for these headers as head pipes connected from the exhaust manifolds to the pipe must be made up anyway. The left stock manifold is actually the only engine component that has to be altered when making the switch. By grinding 1/16 inch off the No. 5 exhaust header and a 45-degree angle on the upper inner edge of the steering box casting, they will just clear. That's the extent of the engine compartment modification.

What about installing a 427 yourself? There's nothing to keep you from it. Only problem is, you won't be allowed to compete in the stock classes — you'll go into Modified Production, Gas or Factory Experimental classes, depending upon which association you compete within. If you only drive it on the street, no problem.

Nickey can supply you with a complete kit, depending on whether you install a 427 right out of a crate or one from a wrecking yard. With a wrecking yard engine, the only parts needed would be the two head pipes connecting the exhaust manifold to the exhaust pipe. This is the only place where welding is needed. The complete swap, except for the head pipes, is a simple bolt in operation. All wearing components are stock Chevy items, no more long distance calls or writing for parts.

It is necessary to use a few of the 350 clutch components with the 427 clutch. In order for the Camaro clutch linkage to work, the clutch fork pivot ball, clutch release bearing, clutch fork and rubber dust cover must be replaced by the 350's corresponding parts. With the engine out and while transposing the clutch parts, Harrell drills a one-eighth inch hole in the top of the left frame motor mount pad for the clutch return

spring. Once the engine is in, it's a bit awkward drilling from underneath the car.

The added weight of the 427 lowers the front of the Camaro approximately one inch. To bring it back to stock height, Harrell installs air conditioning spacers under the front coils. By ordering the car with heavy duty suspension the stiffer springs are strong enough to handle the additional weight so that no other alterations are needed.

Recalling the rear axle wrap-up discovered in our Camaro SS Road Test (December CC), we were curious to see how Harrell overcame the mono-leaf problem. He found that by slightly modifying the Chevy II traction bars, by cutting off the Chevy II shock absorber bolt and replacing it with a 7/16 inch bolt in a new location, he could completely eliminate any wheel hop off the line. The shock then mounts ahead of the bracket, keeping it in the stock location.

The Nickey traction bar replaces the spring plate which attaches the spring and shock absorber to the axle housing and extends forward, stopping just behind the spring hanger. Here it clamps to the spring leaf rather than to the body. Mounted this way, the spring wrap-up is completely eliminated without changing the ride characteristics.

The distributor is the only part to get the super tune. Every-

(continued on page 37)

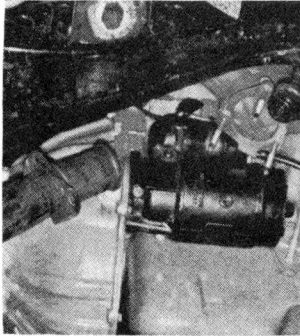
Right side of engine shows an even wider gap between exhaust manifold and head pipe. The 350 starter is used on 427.

Since head pipe to muffler is one piece, Nickey has Ced's Muffler of Chicago, make up new tube to splice into system.

New section of head pipe is slipped over stock tubing and tack welded; then taken loose for finish welding. Left pipe is designed to allow oil filter to remain in original position.

Man . . . like that's a close fit! Although it looks like steering box and manifold touch, there is a 1/16th-inch clearance between the two. Engine torque will pull them farther apart.

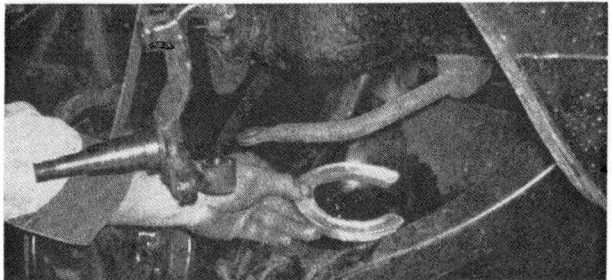

ABOVE — Additional 90 pounds of engine caused front of car to drop one inch. Air conditioner spacer is installed under the front coil springs to bring the Camaro back to the stock height.
RIGHT — Harrell modifies the single point 427 distributor by installing heavy duty points with their greater spring tension. Advance curve is adjusted for each car's specific use.

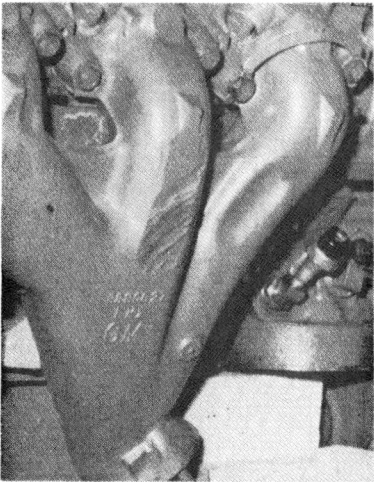

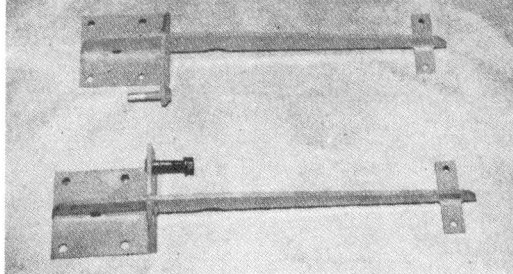

ABOVE—Chevy II traction bars are stocked by speed shops and are a must on the Camaro. The lower unit was converted by shifting the shock mounting bolt.
LEFT — Number five exhaust manifold header is the only engine component to feel the grinder. Heavy casting easily gives up 1/16th-inch with no ill effects.
RIGHT — Steering box gets corresponding "narrowing" by having its upper right corner ground to 45 degrees. That's all that is needed to stop the rattles.

Converted Chevy II traction bars mount in place of the spring plate which attaches the spring and shock absorber to the axle housing. The bars are switched, the right being used on the left and vice versa on the Camaro.
That little piece of rubber weather stripping (arrow) is worth an additional 500 rpm. By using 396 distributor weights and six cylinder Chevrolet springs, Harrell can run the 427 distributor to 8000 rpm without misfire.

TWO CHEVROLET
CAMAROS

THE DILEMMA facing a potential Camaro buyer is a natural one for this time and place in the automotive firmament. This dilemma is particularly perplexing to the buyer who also is an enthusiast, an appreciator of machinery and performance. The problem is not whether to buy the Camaro, but what *kind* of Camaro, for the Camaro probably wears more faces than any other single car now made.

Virtually every part of the car has, or is, an option in some form or another. With the seven engines and nine transmissions listed, there are 19 different power train combinations available. Also, there are two different body styles, coupe and convertible, and even a choice of visible or disappearing headlamps. Exterior colors, vinyl roof coverings (for the sports coupe), paint stripes, interior and exterior trim packages and even wheel/tire options create such a plethora of choices that it is conceivable that a completely custom car could be ordered and another one just like it might never be built. The point is, the Camaro can be almost any kind of car its orderer specifies it to be.

With an eye on this fact, and an eye on the relative performance potential, *CAR LIFE* asked for and received two widely varying Camaros for test purposes. The first was a plain-Jane Six, lightly equipped and low in retail cost. The second was an SS 350, loaded to the drip-moldings with options and accessories, and carrying a pricetag nearly $1000 higher. It should be pointed out that Camaro "goodies" are all extra-cost items, and that the price of ordering that one-only customized car could add up to a somewhat staggering total.

The test Six represented the lower end of the pricing scale and as such emerged quite a bargain in on-the-road performance. With only power steering, radio, white sidewall tires and an appearance group added to the 250-cu. in./155-bhp engine option, it carried an FOB price of only $2791. The engine option, one well worthwhile to any potential customer, adds just $26.35 to the price and measurably increases the car's "drivability."

Chevrolet wisely made the 230-cu. in./140-bhp version of its basic 6-cyl.

SS 350 and Big Six—Both have Virtue and Plenty of Performance

engine the standard engine for Camaro. This represents 30-cu. in. displacement (and 20 bhp) more than the opposition's Mustang Six (200/120)—and it is about on a par with the Barracuda Six 225/145. It also allows Chevrolet to make the next-size-larger Six, the 250/155, available as a low-cost option. The 250/155 is the big Chevrolet's standard engine and, though it is mildly tuned for its workhorse role, it produces enough horsepower to make the Camaro Six an economically interesting performer.

In driving the Big Six Camaro, *CL* found it delivered surprising fuel economy—an actual 19.2 mpg for the 1000-mile test period, with the occasional delivery of over 21 mpg where steady-state driving was encountered. Accelerative performance, with the standard 3-speed all-synchromesh transmission, was entirely adequate, though the column-mounted shift lever and subservient linkage tried to balk every fast shift. (One tester likened it to the Vacuumatic linkage on a 1941 Chevrolet he once owned.) The 30-70

mph passing acceleration time of 10.8 sec. indicates satisfactory freeway on-ramp capability, something that cannot always be said for 6-cyl. cars.

Perhaps the most singularly impressive aspect of the 6-cyl. Camaro was its apparent balance; it had a balanced feeling in driving and cornering, it had a balance in specification, and a balance at the weighing station. Sixes, generally smaller in displacement and size, offer a definite weight advantage over bulkier, but more powerful, V-8s. In the case of the two test Camaros, the 6-cyl. had 210 lb. less weight on the front wheels than did the SS 350, which gave the Six a 54.9%/45.1% front/rear weight distribution to the SS 350's 57.5/42.5%. This balance manifested itself in good cornering and handling characteristics, less rear wheelspin under acceleration, and better braking action. Camaro's inherent understeer was notably less in the Big Six than in the SS 350 version.

The SS 350 really blossoms as a personal/luxury/HP sort of car, though some penalties must be paid

HEADLIGHTS visible is standard style; covered, blank look is the option.

ARMREST handle well is excellent protector, and an unintentional ashtray.

CAMAROS

in both cost and handling characteristics. Acceleration and top speed, of course, are improved with the big 350-cu. in. V-8 underhood, but balance, handling and braking (with the standard drum brakes) are not as good as with the Big Six. However, the SS 350 is likely to be the more popular, simply because it moves with spirit and authority when the throttle pedal is depressed.

Including both engine and trim packages, the SS 350 is readily identifiable by its circumferential "bumblebee" nose stripe. Its special hood has

die-cast simulated louvers, and "SS" emblems appear in numerous strategic locations. Of more interest to the enthusiast buyer is the SS mechanical package: A 350-cu. in. V-8, exclusive to this model, 6-in. wide wheel rims and Firestone Super Sports Wide Oval tires, and firmer spring and shock absorber specification. As a package, this adds $211 to the basic $2572 V-8 coupe price (FOB Detroit), and certainly must be considered something of a bargain.

Another major offering is the "Rally Sport" package. In effect what Detroit

calls an "appearance group," this package is comprised of trim additions and variations. Concealed headlights in a full-width grille are the major features, though special taillights, back-up lights in the rear valance panel, special "RS" emblems and longitudinal paint stripes, and wheel opening and drip rail moldings also are included in the $105 pricetag. The RS group can be ordered with the SS 350 package, for appearance's sake, but the exterior theme is SS rather than RS.

It is in the power train selection that the customer gets the greatest choice. The Sixes, both basic 230-cu. in. and optional 250, provide adequate power for the undemanding driver. The V-8s range from mild to wild, from 302- to 396-cu. in. displacement. There is an engine for virtually every imaginable automotive purpose, including all-out road racing and dragstrip competition. In summary, the power train lineup is:

ENGINES

Displ.	Type	CR	Carb.	Bhp @ Rpm
230	IL-6	8.5	1x1	140 @ 4400
250	IL-6	8.5	1x1	155 @ 4200
302	V-8	11.0	1x4	290 @ 5800
327	V-8	8.8	1x2	210 @ 4600
327	V-8	10.0	1x4	275 @ 4800
350	V-8	10.3	1x4	295 @ 4800
396	V-8	10.25	1x4	325 @ 4800

TRANSMISSIONS/RATIOS

	1st	2nd	3rd	T.C.	Appl.
Man.3-speed	2.85	1.68	1.00	—	6-cyl.
Man.3-speed	2.54	1.50	1.00	—	327, 350
Man.3-speed, HD	2.41	1.57	1.00	—	350, 396
Man.4-speed	3.11	2.20	1.47	—	6-cyl.
Man.4-speed	2.52	1.88	1.47	—	327, 350, 396
Man.4-speed	2.20	1.64	1.27	—	302, 396
Powerglide	1.76	1.00	—	2.40	6-cyl.
Powerglide	1.76	1.00	—	2.10	327, 350
Turbo Hydra-Matic	2.48	1.48	1.00	2.10	396 only

Obviously, the 396/325 is for the big-inch, big-go dragstrip fans, while

SPECIAL INSTRUMENTATION includes fuel, temperature, oil pressure and battery charge gauges, and a clock in console-mounted housing. It's a $79 option.

FILLER CAP is anchored by a cable on SS 350s, is loose on standard models.

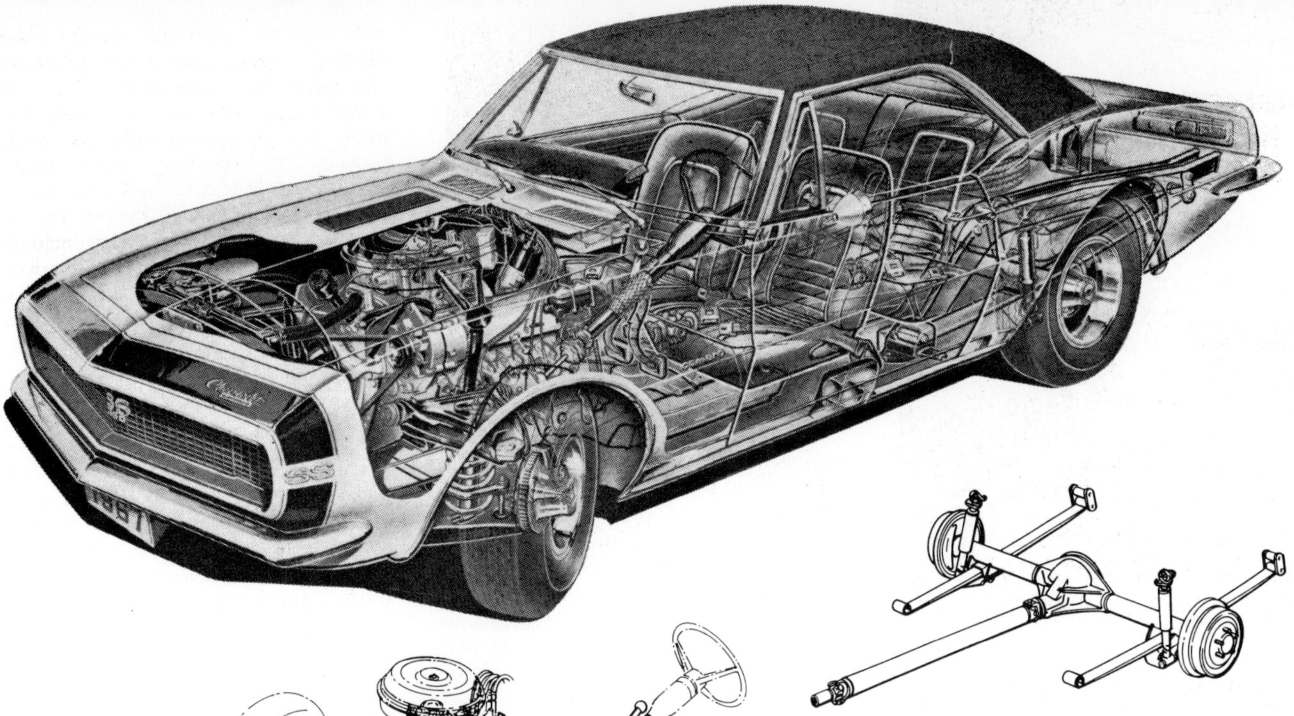

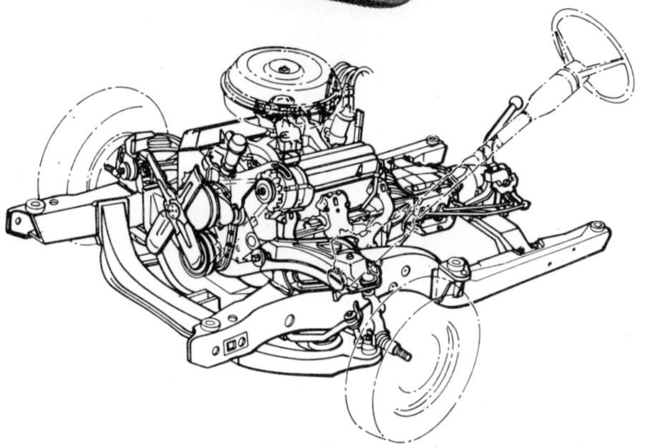

CONSTRUCTION features both a unitized body and a front sub-frame in a "wheelbarrow" arrangement. The box-section sub-frame is rubber isolated from the body, and carries engine, transmission and front suspension. Rear suspension, parallel single-leaf springs, anchors, in rubber, to the body.

the 327/210 is for automatic, air-conditioned, power-assisted, all-around-the-town driving. The 327/275 and 350/295 are well-suited to enthusiastic on- or off-highway use, but the gem of the lot, though it is best-used in maximum performance roles, is the 302, a recent addition to the Chevrolet engine family. It forms the basis of the Z-28 option, a special package of components aimed directly (though GM officially eschews competition) at the sedan class in road racing, where engine displacement is limited to 305 cu. in.

SS 350 HAS deluxe, plastic-chrome-trimmed steering wheel, special horn button as part of its trim package.

HEART OF the SS 350 is the biggest-yet version of Chevrolet's lightweight V-8. Lengthened stroke accounts for increase.

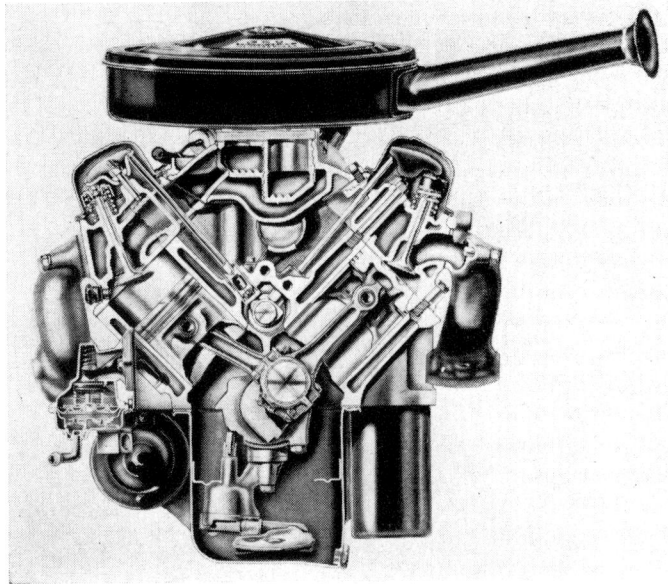

TRUNK space is minimal in Camaros, further restricted by awkward spare tire placement.

The 302 and the 350 are new engine sizes, both exclusive to the Camaro. Both share 327 beginnings and some components. The 302, 327 and 350 all have a 4-in. bore, but the 302 carries the 283 engine's crankshaft, which has a 3-in. stroke; the 327 has a 3.25-in. stroke; and the 350 a 3.48-in. stroke.

Of vast interest to the performance-minded is that power-producing parts from the Corvette's 350-bhp 327 V-8 can be substituted for those on any of the 327s or the 350. This equipment includes a longer duration, higher lift camshaft, (0.447 in. and 342°, including ramps) and mechanical lifters, higher compression (11:1), larger Holley 4-barrel carburetor (800 cfm), and intake manifold, exhaust headers, viscous fan clutch, and big valve (2.02 in. intake, 1.60 exhaust), big port cylinder heads. This is the treatment given a 302 to suit it for racing. Applied to a 350-cu. in. block, like handling ought to produce 375 bhp!

Camaro's unitized body/sub-chassis combination, unique among Chevrolet cars, grew from a minor engineering project that had continued on after the Chevy II debut. Engineers suspected that the concept of a sturdy partial frame in combination with a body structure of near-unitized rigidity offered the most benefit in isolation against road noise and harshness. No particular car—certainly not Camaro—had been envisioned at the time. It was, instead, a rather commonplace example of engineering investigation.

Cobbled Chevy IIs, considered "pre pre-test" cars for the program, were used to try out the various configurations. Though partial frames extending as far aft as the front hangars of the rear spring were tried, it rapidly became apparent the "wheelbarrow" idea was the better compromise. This configuration, later recalled as Camaro work got underway, cantilevered the front stub frame out from mounting points under the seat riser and the cowl base, two to a side. Heavy body bolts, thickly bushed with rubber, were used to secure the mating.

Computer technology, called into play during development of Camaro to a far greater extent than for any other car in General Motors' inventory, speeded up the selection of stub frame parameters. Proving ground tests then bore out the computer's calculations. Isolation of the unitized passenger compartment from shocks and noises of the suspension was at an extremely efficient level.

Front suspension components, quite similar in configuration and geometry to those of the Chevelle, were carried on this deeply-drawn semi-chassis. At the rear, the Chevy II's trouble-free

1967 CHEVROLET
CAMARO SIX

DIMENSIONS

Wheelbase, in.	108.1
Track, f/r, in.	59.0/58.9
Overall length, in.	184.6
width	72.5
height	51.0
Front seat hip room, in.	2 x 20.5
shoulder room	56.7
head room	37.0
pedal-seatback, max.	40.5
Rear seat hip room, in.	54.8
shoulder room	53.8
leg room	30.5
head room	36.7
Door opening width, in.	41.2
Floor to ground height, in.	10.0
Ground clearance, in.	6.3

PRICES

List, FOB factory	$2466
Equipped as tested	2791
Options included: 250/155 engine, emission controls, power steering, radio and appearance groups, tinted windshield, wsw tires.	

CAPACITIES

No. of passengers	5
Luggage space, cu. ft.	8.3
Fuel tank, gal.	18.5
Crankcase, qt.	5.0
Transmission/diff., pt.	3.0/3.5
Radiator coolant, qt.	11.0

CHASSIS/SUSPENSION

Frame: Unitized body; front sub-frame.
Front suspension type: Independent by s.l.a., coil springs, telescopic shock absorbers, ball-joint steering spindles, antiroll stabilizer.

ride rate at wheel, lb./in.	124
anti-roll bar dia., in.	0.6875

Rear suspension type: Live axle, Hotchkiss drive; single-leaf parallel springs, telescopic shock absorbers.

ride rate at wheel, lb./in.	121

Steering system: Coaxial, power assisted semi-reversible recirculating ball nut; parallelogram linkage.

gear ratio	15.6
overall ratio	17.5
turns, lock to lock	3.0
turning circle, ft. curb-curb	37.0
Curb weight, lb.	2998
Test weight	3408
Weight distribution, % f/r	54.9/45.1

BRAKES

Type: 2-circuit hydraulic; self-adjusting duo-servo shoes in composite drums.

Front drum, dia. x width, in.	9.5 x 2.5
Rear drum, dia. x width	9.5 x 2.0
total swept area, sq. in.	268.6

Power assist: Integral, vacuum.

line psi @ 100 lb. pedal	n.s.

WHEELS/TIRES

Wheel size	14 x 5J
optional size available	14 x 6JK
bolt no./circle dia., in.	5/4.75

Tires: B.F. Goodrich Silvertown 660

size	7.35-14
recommended inflation, psi	24
capacity rating, total lb.	4640

ENGINE

Type, no. cyl.	ohv, IL-6
Bore x stroke, in.	3.875 x 3.53
Displacement, cu. in.	250.174
Compression ratio	8.5
Rated bhp @ rpm	155 @ 4200
equivalent mph	104
Rated torque @ rpm	235 @ 1600
equivalent mph	40
Carburetion	Rochester, 1x1
barrel dia., pri./sec.	1.56

Valve operation: Hydraulic lifters, pushrods, overhead rockers.

valve dia., int./exh.	1.72/1.50
lift, int./exh.	0.388/0.388
timing, deg.	62-94, 92-63
duration, int./exh.	336/336
opening overlap	125

Exhaust system: Single reverse-flow muffler.

pipe dia., exh./tail	2.0/2.0

Lubrication pump type: gear

normal press. @ rpm.30-45 @ 1500	

Electrical supply: alternator

ampere rating	37 @ 12 V.
Battery, plates/amp. rating	54/45

DRIVE TRAIN

Clutch type: Diaphragm; single disc.

dia., in.	9.12

Transmission type: Manual 3-speed.

Gear ratio 4th ( ) overall	
3rd (1.00)	3.31
2nd (1.68)	5.56
1st (2.85)	9.44

synchronous meshing?...3 forward
Shift lever location...steering column
Differential type: Hypoid; overhung pinion.

axle ratio	3.31

single-leaf springs (shortened to 56 in.) and Hotchkiss drive were specified.

Using a program devised from years of test data compilations, the computer told Chevrolet engineers how to piece it all together to provide optimum handling qualities. Spring rates and roll couple distribution were determined by computer. Electronic guidance also suggested proper stiffness for the various rubber bushings throughout the chassis, the vertical positioning of the rear shock absorbers, and the angular placement of the rear springs to make room for a larger fuel tank. As chassis engineer Charles Rubly explained it, "You can predict in advance that you're going to have a car you're going to like or not like. It saved us an awful lot of cut-and-try." Using the computer to speed up engineering calculations, he continued, meant that "you have arrived at the best you're going to be able to do with that vehicle, without changing parameters."

Still, some troubles wait until proving ground tests materialize. One such with Camaro was what Rubly called "a jitter problem." It was caused, it turned out, by a type of rear spring eye that limited fore and aft flexibility. Rear axle tramp was another area that

demanded attention once the 350-cu. in. engine moved into the car. With smaller Sixes and the 327 V-8s, the problem was not so apparent. To control this deficiency, a single traction bar was installed on the right side as part of the package for all V-8s with manual 4-speed transmission.

Only one notable exercise was involved in the body engineering. The convertible model, lacking the strength of a rigid roof panel, had to regain beam strength with a double channel insert inside the length of the rocker sills. Then, because too much rigidity would adversely affect riding qualities, hydromechanical vibration dampers were installed at each corner of the car, inside the trunk and engine compartments. These small barrel-shaped cylinders with a spring-mounted weight suspended in an oil bath damp out cyclic motions induced in the body from some road surfaces.

The computer was called in again during one stage of body design and production. The stylists' shape was put on tape which then, through a computer, operated a machine which cut die models. "It was almost perfect," reported General Manager E.M. (Pete) Estes, "probably more perfect than can be done normally." Camaro's

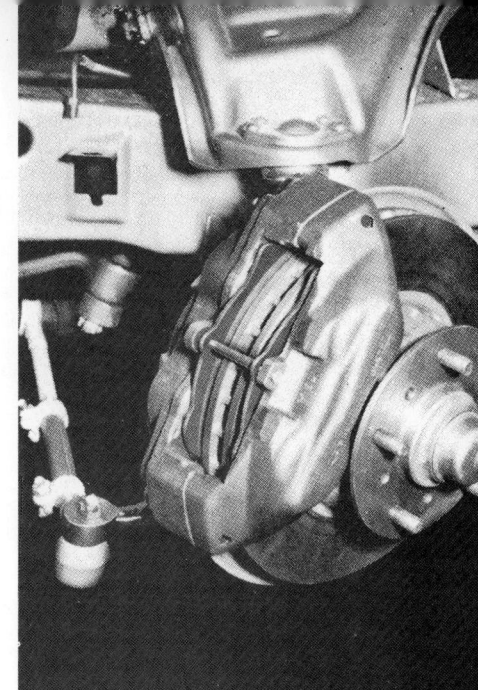

DISC brake option uses radially-vented rotors and four pistons per caliper.

shape and contours, it might be noted, originated within the Corvette styling studio, which accounts for such similarities as the extreme tuck-under at the lower side panels, the sweeping

CAR LIFE ROAD TEST

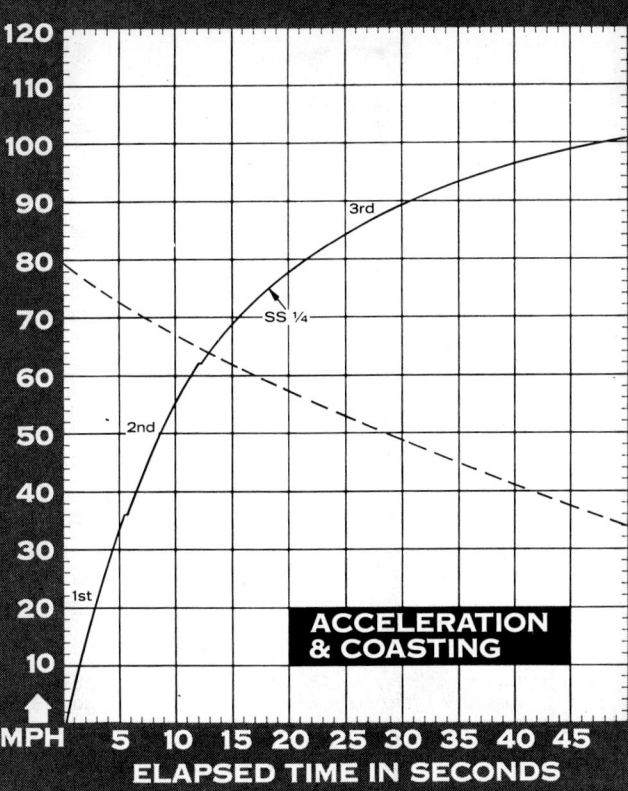

ACCELERATION & COASTING

ELAPSED TIME IN SECONDS

CALCULATED DATA

Lb./bhp (test weight)	22.0
Cu.ft./ton mile	103.4
Mph/1000 rpm (high gear)	24.7
Engine revs/mile (60 mph)	2430
Piston travel, ft./mile	1430
Car Life wear index	34.7
Frontal area, sq. ft.	20.5
Box volume, cu. ft.	395

SPEEDOMETER ERROR

30 mph, actual	32.0
40 mph	41.7
50 mph	52.1
60 mph	62.7
70 mph	73.6
80 mph	85.6
90 mph	94.8

MAINTENANCE INTERVALS

Oil change, engine, miles	6000
transmission/dif.	as req.
Oil filter change	6000
Air cleaner service, mo.	6
Chassis lubrication	6000
Wheelbearing re-packing	as req.
Universal joint service	none
Coolant change, mo.	24

TUNE-UP-DATA

Spark plugs	AC 46N
gap, in.	0.033-0.038
Spark setting, deg./idle rpm	6/500
cent. max. adv., deg./rpm	28/2800
vac. max. adv., deg./in. Hg.	21/14.5
Breaker gap, in.	0.019
cam dwell angle	31-34
arm tension, oz.	19-23
Tappet clearance, int./exh.	0/0
Fuel pump pressure, psi	3.5-4.5
Radiator cap relief press., psi	15

PERFORMANCE

Top speed (4200), mph	104
Shifts (rpm) @ mph, manual	
3rd to 4th ( )	
2nd to 3rd (4200)	62
1st to 2nd (4200)	36

ACCELERATION

0-30 mph, sec.	4.8
0-40 mph	6.2
0-50 mph	8.7
0-60 mph	11.4
0-70 mph	15.6
0-80 mph	21.7
0-90 mph	31.0
0-100 mph	
Standing ¼-mile, sec.	18.5
speed at end, mph	75
Passing, 30-70 mph, sec.	10.8

BRAKING

(Maximum deceleration rate achieved from 80 mph)

1st stop, ft./sec./sec.	21
fade evident?	no
2nd stop, ft./sec./sec.	22
fade evident?	slight

FUEL CONSUMPTION

Test conditions, mpg	19.2
Normal cond., mpg	19-22
Cruising range, miles	351-407

GRADABILITY

4th, % grade @ mph	
3rd	12 @ 52
2nd	20 @ 43
1st	26 @ 32

DRAG FACTOR

Total drag @ 60 mph, lb	136

ENGINE COMPARTMENT of SS 350 is filled by engine, though there's enough space left for power accessories, air conditioning.

fender line, and the tenacious look of big black tires jutting from all corners.

Part of that look, however, resulted from a late change in specifications. The rear tread was widened to 58.9 in.

well after the computer had laid down the ideal dimensions, but the effect, as any hot rodder knows, only enhanced the handling qualities. The completed car stands on an excep-

tional tread-to-wheelbase ratio. And, before stylists finished with the contours, the clay model was put through wind tunnel tests to prove the curves were aerodynamically slippery. Though this caused a few minor changes in a few odd corners of the sheet metal, it also established the functionality of body styling.

The Camaro carries the same 9.5-in. brake drums as do Corvair, Chevy II, and Chevelle. They are not much more than adequate on the Camaro, just as they were on other Chevrolet products *CL* has tested. Fortunately for the enthusiast driver, Chevrolet now has two interesting, and superior, options: Disc front brakes and sintered metallic-lined shoes in the standard-sized drums. Both can be power assisted. The disc/drum system costs $79, the metallic linings $37; the power booster, which should be a must for the metallics, adds another $42. *CL*'s recommendations are: The disc system, power assisted or not, for everyday driving, and the boosted metallics for dragstrip type of competition. For even tougher usage, Chevrolet includes a special combination in the Z-28 option: Disc front brakes with heavy-duty front pads, and metallic linings in the rear drums, all power assisted.

1967 CHEVROLET
CAMARO SS 350

DIMENSIONS

Wheelbase, in.	108.1
Track, f/r, in.	59.0/58.9
Overall length, in.	184.6
width	72.5
height	51.0
Front seat hip room, in.	2 x 20.5
shoulder room	56.7
head room	37.0
pedal-seatback, max.	40.5
Rear seat hip room, in.	54.8
shoulder room	53.8
leg room	30.5
head room	36.7
Door opening width, in.	41.2
Floor to ground height, in.	10.0
Ground clearance, in.	6.3

PRICES

List, FOB factory............$2572
Equipped as tested...........3630
Options included: SS 350, RS pkg., emission controls, power steering and brakes, 4-speed trans., radio and rear antenna, custom interior, instrument group, Positraction, vinyl roof cover, tinted windshield.

CAPACITIES

No. of passengers	5
Luggage space, cu. ft.	8.3
Fuel tank, gal.	18.5
Crankcase, qt.	5.0
Transmission/diff., pt.	3.0/4.0
Radiator coolant, qt.	15.0

CHASSIS/SUSPENSION

Frame: Unitized body; front subframe.
Front suspension type: Independent by s.l.a., coil springs, telescopic shock absorbers, ball-joint steering spindles.
ride rate at wheel, lb./in. ...125
anti-roll bar dia., in. ...0.6875
Rear suspension type: Live axle, Hotchkiss drive; single-leaf parallel springs, telescopic shock absorbers.
ride rate at wheel, lb./in. ...125
Steering system: Coaxial power assisted, semi-reversible recirculating ball nut gear; parallelogram linkage.
gear ratio ...15.6
overall ratio ...17.5
turns, lock to lock ...3.0
turning circle, ft. curb-curb ...37.0
Curb weight, lb. ...3210
Test weight ...3620
Weight distribution, % f/r. 57.5/42.5

BRAKES

Type: 2-circuit hydraulic with tandem master cylinders; self-adjusting duo-servo shoes in composite drums.
Front drum, dia. x width, in. 9.5 x 2.5
Rear drum, dia. x width ...9.5 x 2.0
total swept area, sq. in. ...268.6
Power assist: Integral, vacuum.
line psi @ 100 lb. pedal. ...n.s.

WHEELS/TIRES

Wheel size ...14 x 6JK
optional size available ...14 x 5J
bolt no./circle dia. ...5/4.75
Tires: U.S. Royal High Performance
size ...7.35-14
recommended inflation, psi ...26
capacity rating, total lb. ...4840

ENGINE

Type, no. cyl.	ohv, 90° V-8
Bore x stroke, in.	4.00 x 3.48
Displacement, cu. in.	349.670
Compression ratio	10.5
Rated bhp @ rpm	295 @ 4800
equivalent mph	111
Rated torque @ rpm	380 @ 3200
equivalent mph	74
Carburetion	Rochester, 1x4
barrel dia., pri./sec.	1.38/2.25

Valve operation: Hydraulic lifters, pushrods, overhead rockers.
valve dia., int./exh. ...1.94/1.50
lift, int./exh. ...0.390/0.410
timing, deg. ...36-94, 86-54
duration, int./exh. ...310/320
opening overlap ...90
Exhaust system: Dual, with resonators.
pipe dia., exh./tail. ...2.50/2.00
Lubrication pump type ...gear
normal press. @ rpm. 30-45 @ 1500
Electrical supply ...alternator
ampere rating ...37 @ 12 V.
Battery, plates/amp. rating ...66/61

DRIVE TRAIN

Clutch type: Diaphragm; semi-centrifugal disc.
dia., in. ...11.0
Transmission type: Manual 4-speed.

Gear ratio		
4th (1.00) overall	3.55
3rd (1.47)	5.17
2nd (1.88)	6.62
1st (2.52)	8.87

synchronous meshing? ...4 forward
Shift lever location ...floor console
Differential type: Hypoid; overhung pinion.
axle ratio ...3.55

Suspension stiffnesses vary with the engine specification, though all are definitely firm. The basic Six has ride rates of 124 lb./in. in front and 121 lb./in. at rear, while the SS 350 has 125 and 131, front and rear. All Camaros carry a 0.687-in. front antiroll stabilizer, and the overall roll resistance is notably firm. The Camaro corners in a relatively flat attitude, without that annoying front-end tucking that is so apparent in other cars. The feeling, despite a quick but quite insensitive power steering, is that the car is inordinately nimble.

Inspection of the two test vehicles revealed little to really carp about. Exterior panels were slightly wavy and their paint finish had a few areas of "orange peel." But the panels fit together well in their intended pattern. The interiors appeared made well enough, though the plain model looked better than the chrome-splotched SS 350. Reflections bounced irritatingly off the SS's steering wheel and A pillars when driven in the bright sunshine. Two other specific, if minor, points of complaint were: Odd refraction by cone-shaped plastic instrument covers, and sharp edges on the 4-speed's shift lever which gouged the driver's hand during gear-changing.

BIG SIX engine takes up less space, has greater access for servicing. Performance of this engine is surprisingly good for only 250 cu. in.

"Fun-to-drive" keynotes the Camaro. CL testers can't remember when they've had cars with such a high Fun Factor. Evidently it works that way for others, too. The editor of another automotive magazine, which gave its yearly award to another car, bought an SS 350 for his own personal use. The reason? "Because it's fun to drive," he said. ∎

CAR LIFE ROAD TEST

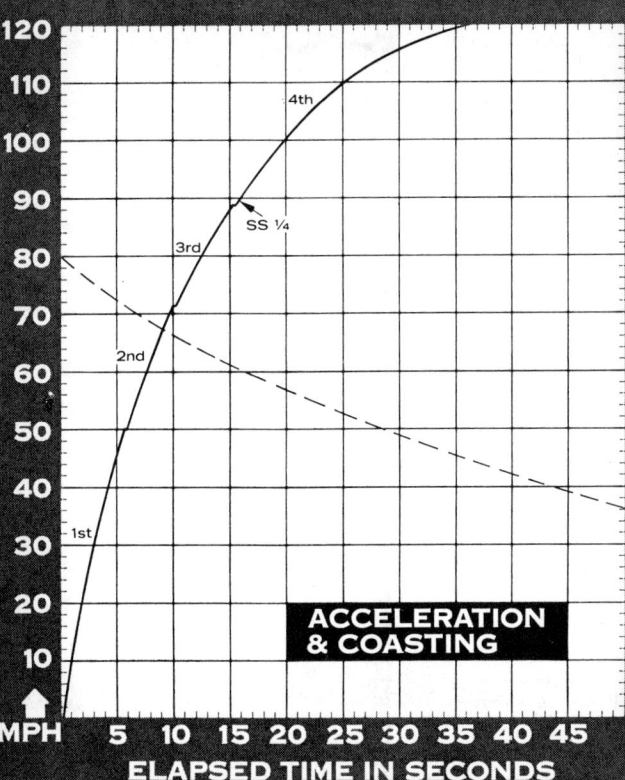

ACCELERATION & COASTING

ELAPSED TIME IN SECONDS

CALCULATED DATA

Lb./bhp (test weight)	12.3
Cu. ft./ton mile	146
Mph/1000 rpm (high gear)	23.2
Engine revs/mile (60 mph)	2590
Piston travel, ft./mile	1500
Car Life wear index	38.9
Frontal area, sq. ft.	20.5
Box volume, cu. ft.	395

SPEEDOMETER ERROR

30 mph, actual	30.6
40 mph	44.3
50 mph	52.3
60 mph	61.6
70 mph	72.0
80 mph	81.8
90 mph	92.7

MAINTENANCE INTERVALS

Oil change, engine, miles	6000
trans./dif.	as req.
Oil filter change	6000
Air cleaner service, mo.	6
Chassis lubrication	6000
Wheelbearing re-packing	as req.
Universal joint service	none
Coolant change, mo.	24

TUNE-UP DATA

Spark plugs	AC 44
gap, in.	0.033-0.038
Spark setting, deg./idle rpm	8/500
cent. max. adv., deg./rpm	26/4100
vac. max. adv., deg./in. Hg	15/15.5
Breaker gap, in.	0.019
cam dwell angle	28-32
arm tension, oz.	19-23
Tappet clearance, int./exh.	0/0
Fuel pump pressure, psi	5.25-6.50
Radiator cap relief press., psi	15

PERFORMANCE

Top speed (5200), mph	120
Shifts (rpm) @ mph, manual	
3rd to 4th (5500)	89
2nd to 3rd (5500)	71
1st to 2nd (5500)	50

ACCELERATION

0-30 mph, sec.	2.9
0-40 mph	4.2
0-50 mph	5.8
0-60 mph	7.8
0-70 mph	10.0
0-80 mph	12.8
0-90 mph	16.1
0-100 mph	19.8
Standing ¼-mile, sec.	15.8
speed at end, mph	89
Passing, 30-70 mph, sec.	7.1

BRAKING

(Maximum deceleration rate achieved from 80 mph)

1st stop, ft./sec./sec.	21
fade evident?	slight
2nd stop, ft./sec./sec.	22
fade evident?	yes

FUEL CONSUMPTION

Test conditions, mpg	16.6
Normal cond., mpg	16-19
Cruising range, miles	296-352

GRADABILITY

4th, % grade @ mph	19 @ 74
3rd	25 @ 63
2nd	32 @ 53
1st	41 @ 36

DRAG FACTOR

Total drag @ 60 mph, lb.	128

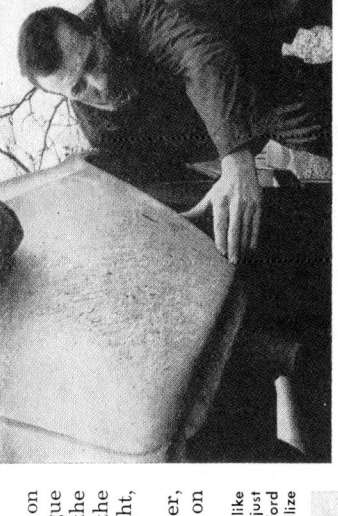

COMFORT, particularly on long trips, was rated high by Camaro owners, many reporting the bucket seats as being the most comfortable they had ever sat in. A few didn't like the pushbutton seat feature

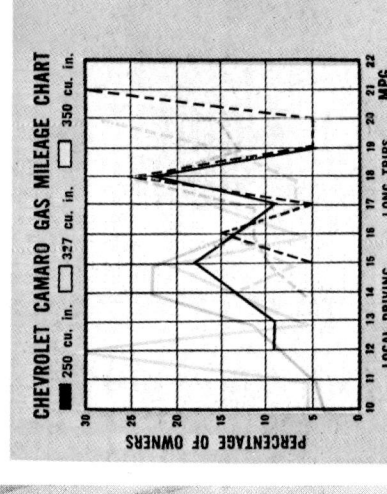

CHEVROLET CAMARO GAS MILEAGE CHART

250 cu. in.   327 cu. in.   350 cu. in.

PERCENTAGE OF OWNERS / MPG — LOCAL DRIVING --- LONG TRIPS

zling example of our highly technological times, yet I should think Mr. McPherson and his colleagues might ponder such all-too-human comments as: "Rear end hops and winds up like a yo-yo under normal acceleration." —Missouri draftsman.

"Traction from a dead stop is nonexistent." —Iowa grocery manager.

Complaints about the rear-suspension setup lead the list of Camaro-owner dislikes, "earning" a surprisingly high Frequency-of-Mention Rating (FMR) of 31.8 percent. To give you an idea of how many Camaro owners commented on this aspect of the car, the next most disliked feature—lack of rear-seat legroom—garnered an FMR of only 15 percent.

Camaro owners who commented on the car's behavior aft were often vague in expressing just what it is about the rear suspension they don't like, but the overall thought is that it's too light, "flimsy" and "skittish."

Somewhat paradoxically, however, owners placed handling second on

INBOARD TURN of rear fenders was deplored by many owners, saying it was a dirt and mud catcher that made it hard to keep the car clean. A few reported chipped finish as a result of wheel-thrown gravel

STINGY TRUNK SPACE was ranked fourth on dislike list, several owners feeling Chevy had cut it just a bit too fine. Associate auto editor Bill Hartford reported spare tire made space awkward to utilize

PM OWNERS REPORT

CHEVROLET CAMARO

Computer analysis played a key role in designing Chevy's new sportster, but Camaro owners report just enough went awry to prove these electronic marvels aren't foolproof

By BILL KILPATRICK, Auto Editor

"CHEVROLET'S NEW CAMARO, perhaps more than any other of today's cars, is a true product of the computer age."

So says a recent press handout. But if the comments of many Camaro owners responding to PM's survey accurately mirror the results of this computerizing, there seems to have been a short circuit in one department.

According to the press release that quotes the Chevy chief engineer for passenger cars, Donald H. McPherson, computers analyzing the rear suspension for the Camaro indicated that "by moving the shock absorbers outboard of the springs and mounting them nearly vertical, instead of the usual diagonal arrangement, the ability of the wheels to more closely follow and maintain contact with washboard road surfaces and during cornering was much improved."

All very well, and no doubt a daz-

their list of Camaro praises, affording it an FMR of 62.9 percent. The conclusion would seem to be that suspension—like beauty—is in the eye of the beholder.

What owners surveyed by PM were lavish with was praise for the Camaro's looks and overall styling, heaping on the car's sporty appearance an FMR of 65.3 percent. Yet even here there was a pointed negative note: Many owners (an FMR of 14 percent) were quite outspoken in their disapproval of the rather severe inboard curvature of the rear fenders, claiming it resulted in a constant spray of mud and road dirt being thrown up by the tires.

Another styling regret expressed by several Camaro owners is that Pontiac's Firebird had stolen their thunder; they had hoped the styling would remain strictly Chevy and Camaro.

A summary of the likes, dislikes and general driving and ownership experiences of Camaro owners surveyed by PM appears below. The percentages listed reflect the frequency with which owners mentioned specific items. The summary is based on 640,163 miles of

CHEVROLET'S NEW CAMARO

driving—around town and on long trips. Comments pertaining to Camaro owner likes and dislikes are quoted below, again in order of frequency mentioned. The boldface asides are mine.

As pointed out above, what owners are most happy with is the car's styling.

"Fell in love with the looks of it."—Delaware secretary.

"Eye-catching. Just plain sharp."—Indiana steelworker.

Owners like the way the Camaro handles, too. Even some of those who weren't too happy with the rear suspension had kind words for other handling aspects.

"Good maneuverability for city driving."—Iowa radio news director.

"Handles great at high speeds."—Mississippi college student.

Next on the list of praises were orchids for the Camaro's get up and go.

"Excellent passing pickup."—Arizona physician.

"Lots of pep, even with a Six."—Maine mechanic.

"Snappy takeoff."—Florida, retired.

▶ **This gentleman is 75 years old. Oh you kid!**

In addition to praises for the Camaro's scoot, owners have nice things to say about the car's all-round performance.

"Everything—power, economy, quietness—goes together nicely."—Michigan lathe operator.

"Best performer on the road."—Ohio lab technician.

Ranked fifth on the list of owner likes was the Camaro's economy.

"Fuel economy exceeded my expectations."—Maryland electronics technician.

"Mileage with the Camaro is better than with the '65 Fury I traded in."—Kansas supervisor.

Switching to the minus side of the owner ledger, we've already cited appropriate owner complaints about the rear-end engineering. Next in sour volume were beefs about rear-seat legroom.

"I wouldn't torture anyone by making them sit in back for any length of time."—Montana, student.

"Very little legroom."—Connecticut bacteriologist.

▶ **Yeah, for very little legs. And that's all. The car's not designed as a multipassenger carrier.**

Some of the sharpest owner criticism of the Camaro appeared in comments about the inboard curvature of the rear fenders.

"Extension of the rear wheels beyond the wheel housing makes mud flaps necessary."—Massachusetts student.

"Rear wheels throw gravel, mud, dirt and so on up against the fenders, chipping the paint."—Michigan factory worker.

Comments pertaining to the fourth-ranked item on the dislike list can be summed up simply by reporting that owners who mentioned it thought the trunk was much too small. Yet here again one is tempted to point out that a big trunk—like lots of rear-seat legroom—isn't the name of Camaro's game.

In fifth place on the gripe list were some salty comments about rattles.

"Rattles pretty bad for a new car."—Indiana assembly worker.

"Rattles more after four months than my Corvair did after three years."—Maryland computer programmer.

Back on the plus side, Camaro owners had nice things to say about the car's ride ("Smooth."—Louisiana secretary) ("Rides like a car costing much more."—Massachusetts truck driver) and overall size ("Just right for my needs."—Ohio plasterer) ("Perfectly proportioned for the type of car it is."—Colorado engineer). Camaro owners also praised the car's all-round comfort, its "feel" on the road, and its interior styling.

Winding up the complaint list, many Camaro owners squawked about poor gas mileage ("Mileage should be better for a car this small."—Iowa salesman), the seemingly ever-present poor workmanship ("Bolts were loose, doors improperly adjusted, paint was bad, upholstery job was sloppy."—Minnesota locomotive engineer), restricted rear visibility ("Vision limited through rear window."—Alabama, U.S. Navy), wind noise, and—as reported by those who own the SS model—a certain amount of trouble with the "hide away" headlight mechanism.

Perhaps because the Camaro is an all-new car, PM's survey of its owners elicited rather positive, even colorful, comments. Apparently, the concept of a "personal" car involves personal feelings.

"I like everything about my Camaro."—Colorado receptionist.

"It's a dream."—Kansas entertainer.

"We're trading it for something else . . . anything."—Minnesota technician.

"All in all, General Motors has made itself an Edsel."—Mississippi student.

Harking back to our opening illustration of computers used in the design and engineering of the Camaro perhaps going somewhat awry (at least in many owner opinions), a Missouri draftsman said:

"Yes, I know it was engineered by computer, and after owning the results, I feel I don't have to worry about being replaced by pushbuttons."

A comforting thought, that.   ★ ★ ★

CRAMPED REAR-SEAT LEGROOM "earned" many owner brickbats, was ranked second on dislike list. Yet Camaro was not designed to be a true four-passenger car. At best, seat is only short-hop convenience

327-CU.-IN. ENGINE with 2-bbl. carburetor was most popular with Camaro owners responding to PM survey. Reports of engine trouble were few and most owners thought engine cranked out plenty of power

Summary of Camaro Owners Reports*

Total miles driven ... 640,163

Average mpg:
- 250-cu.-in. engine ... 17.5
- 327-cu.-in. engine ... 16.0
- 350-cu.-in. engine ... 16.0

Specific likes:
- Styling ... 65.3%
- Handling ... 62.9
- Power/Pickup ... 29.8
- Performance overall ... 21.8
- Economy ... 16.1
- Ride ... 12.1
- Size ... 10.5
- Comfort ... 10.5
- "Roadability" ... 10.7
- Interior styling ... 8.1

Specific dislikes:
- Rear suspension ... 31.8%
- Rear-seat legroom ... 15.0
- Rear fender design ... 14.0
- Trunk too small ... 11.2
- Rattles ... 9.3
- Poor workmanship ... 9.3
- Poor rear visibility ... 6.5
- Wind noise ... 6.5
- Headlights "stick" ... 5.6

Mechanical troubles?
- Yes ... 60.3%
- No ... 39.7

What kind of trouble?
- Headlight mechanism ... 32.9%
- Transmission ... 16.5
- Brakes ... 10.1
- Carburetor ... 7.6
- Clutch ... 6.3
- Speedometer cable ... 6.3

Dealer service satisfactory?
- Yes ... 52.6%
- No ... 38.2
- "Partly" ... 3.9
- "Not back yet" ... 5.3

Why the Camaro?
- Style ... 70.5%
- Size ... 21.7
- "Different" ... 10.9
- Handling ... 10.9
- Past Chevy experience ... 9.3
- Price ... 7.8
- GM reputation ... 8.9
- Economy ... 6.2

Power options:
- "None" ... 50.5%
- Steering ... 46.4
- Brakes ... 14.4
- Seats ... 3.1
- Windows ... 1.0

Camaro your only car?
- Yes ... 42.6%
- No (own two) ... 40.3
- No (own three or more) ... 17.1

Other cars owned:
- Chevrolet ... 45.9%
- Oldsmobile ... 12.2
- Ford ... 8.1
- Pontiac ... 8.1
- Cadillac ... 6.8
- Volkswagen ... 6.8
- Chevy pickup ... 6.8
- Corvair ... 4.1
- Chrysler ... 4.1
- Ford pickup ... 4.1

Age distribution of owners:
- Under 20 ... 12.2%
- 20-24 ... 25.2
- 25-29 ... 18.9
- 30-34 ... 8.4
- 35-44 ... 6.9
- 45-49 ... 7.6
- 50-54 ... 8.4
- 55-59 ... 3.8
- 60-64 ... 1.5
- 65 and over ... 0.8

*Where applicable, percentages may not equal 100 percent due to rounding and/or insufficient sample.

427 Camaro

BY MARTYN L. SCHORR

THE STREET ROD IS DEAD— LONG LIVE THE SHOWROOM SPECIALTY CAR

EARLY-MODEL Olds and Pontiac coupes with late-model GM engines, Fords and Studes with Caddy and Chrysler hemi mills, and Deuce coupes have become just about extinct on the street scene. No self-respecting rodder would even associate himself with the above mentioned ex-hot street setups. They're Outsville!

The street play today is a sophisticated show-go scene. A place where you can see more interesting machinery than you can at a rod and custom show. The street scene has become the *in* scene.

The big street rebellion is in effect on both coasts. Enthusiasts are *again* sinking big bucks into street wheels, shying away from all-out competition machines. And the smart ones are going the double-threat route, making it big on the strip in the stock classes and even bigger in the Drive-In Eliminations!

Partially responsible for the upsurge in street interest are the Detroit iron-makers. Anyone can walk into a showroom, plunk down a reasonable down payment and ride out in a

NEW-BREED SUPER-HYBRIDS TAKE OVER THE STREET SCENE

supercar that'll romp from 0 to 60 in the low 7's and tour the circuit in the mid-14's with trap speeds close to 100 mph. Then all he has to do is add headers, the right gears, beef up the suspension, add cheater tires and go for a fine tune job at the local dyno shop and he has a stock functional machine that's capable of sucking a hot setup swap-mobile of yesteryear up one pipe and blowing it out the other.

Another popular treatment for the street this year is the conversion of a Super/Stock racer into tractable shape. Even if you run a stock consumer engine under the hood you are sure to scare away everybody at a light as long as you are running a jacked suspension, big tires, plexi windows, glass or aluminum panels and a scooped hood. Some guys are even running straight drag-style tube front ends and altered wheelbase chassis on the street these days. Anything goes on the street and you don't have to build to satisfy any tech inspector!

The really big threat on the street scene today is the hot-rodded super specialty stocker. Cars that fall into this class are 427 (converted from 428) Shelby GT Mustangs, Ram Air Pontiac Firebirds and 427 Camaros. These machines are just about the hottest on the street today. Even the converted Super/Stocks shy away from these hybrid beasts. Just about the hottest in this class are the various 427 Camaros which, when even slightly modified (headers, gears, tires, etc.), are capable of going to 60 from a standing start in 5 seconds.

Because of the renewed interest in street machinery we scoured the New York City suburbs looking for the most feared and most desired of the new-breed super specialty street machines. As expected, we found nothing quicker or faster than some select 427 Camaros. Even though this is not a full factory production vehicle and it does not carry the *full* 5 & 50 factory warranty (90 days on engine, 5&50 on running gear and car), the 427 Camaro has caught on with the non-racing-oriented street enthusiasts.

Interested in finding out exactly how far the purchaser of one of these hybrids goes in the line of performance boosting goodies, we checked out an SS-427 Camaro owned by Bruce Levinson, a typical, young married family man living in Manhasset, New York.

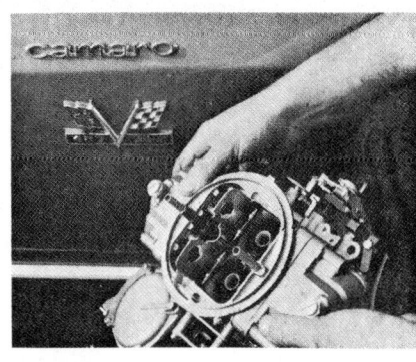

Hot setup Holley three-barrel carb flows as much air as factory tri-power assembly. Four-tube lightweight tuned headers are made by Bill Thomas for 427 Camaro specials.

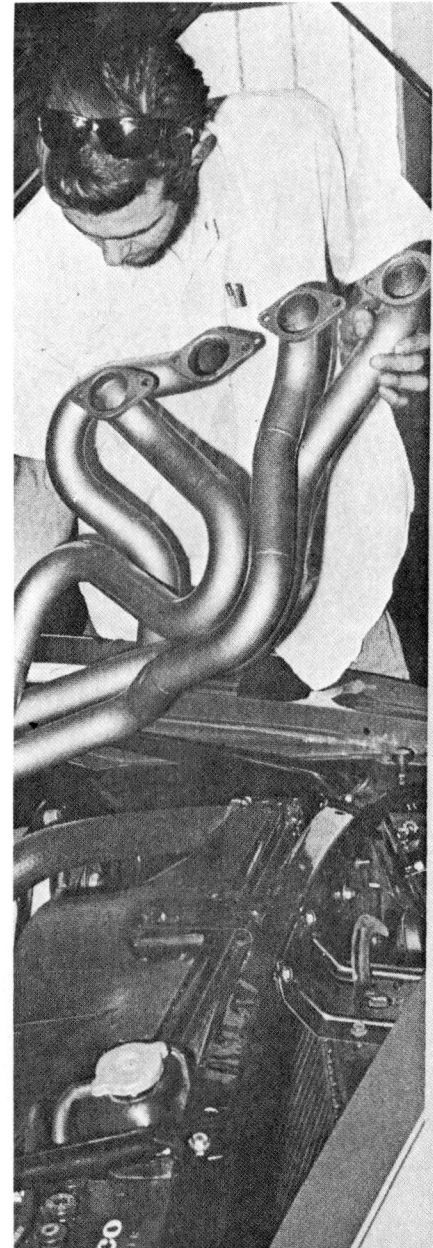

Wood wheel jazzes up optional interior. Buckets are stock.

Head-turning 427 Camaro boasts mags all around, electric blue paint, a vinyl roof and hood pins. Car is unbeatable on the street.

Goodyear Wide Track shoes on mags improve road and track handling. Suspension is beefed. With three-barrel 435-hp plus mill Camaro is quicker and faster than hottest 'Vettes.

Bruce had gone the typical '55 Chevy street-strip route and was tired of finding carbon copies everywhere he went. He was also tired of being labled "hot rodder" every time he rolled up to a light, or when he took his wife shopping. "I was tired of being prodded into a stoplight Grand Prix every time I rolled up to a red light. If the driver in the next lane was a go-fast type, he would sound his pipes the minute he eyeballed my mag wheels and jacked suspension," said Bruce. "With the SS-427 Camaro, very few guys bother me and I know that I have enough blast in reserve to handle *any* situation," he added.

Bruce Levinson's 427 Camaro represents a sizeable investment because of all the extra equipment he

Bill Thomas traction bars aid axle control, prevent spring windup. Cure Ride shocks are recommended for 427 Camaros.

added to the basic 427 package. The car was purchased from Baldwin Chevrolet in Baldwin, New York, and prepared for Bruce by the designer of the SS-427 Camaro, Joel Rosen at Motion Performance, Inc., also in Baldwin. Because Bruce was interested in strip as well as street operation, many performance-plus goodies were added by Motion Performance before the car was delivered. Since these modifications are considered standard by the ma-

jority of the go-fast set, we'll go into detail to give you a better idea of what's the hot setup.

The standard SS-427 Camaro comes through with a 425-hp motor featuring a single Holley quad, a good solid-lifter camshaft and tuned cast-iron headers. The engine is basically the '66 427 cubic-inch 425-hp Corvette top line powerhouse which is no longer offered in the '67 Corvette option book. Bruce ordered his car with the '67 Corvette 435-hp tri-power motor, then after the car was delivered changed his mind in favor of a single quad setup. He traded in the tri-power for an L-88 Corvette racing intake package with Vic Edelbrock's new three-barrel Holly carb. This manifold, which is available from Chevrolet, has a hogged-out riser and is of lightweight aluminum construction. The Edelbrock three-barrel Holley quad flows as much as the tri-power manifold, offers more response throughout the range and is simpler to maintain. This is a very popular modification with the Chevy big-block set.

After the intake system was swapped, Bruce decided to dump the stock cast-iron headers in favor of a set of Bill Thomas four-port tuned lightweight tube steel headers which can be uncorked for strip running. The headers were installed and the dual exhaust system modified to allow a simple swap from street to strip for Sunday afternoon blasting.

Because the 427-cube engine is good for 7000-rpm shifts in stock condition and weekend drag blasts were on the plans board, Bruce decided to trade the super-beefy factory clutch package in for a Schiefer explosion-proof clutch-flywheel assembly. And for added foot insurance he added a steel scattershield bellhousing.

Included in the Schiefer conversion package are an aluminum flywheel, Super Rev Lok diaphragm-style pressure plate and a high performance disc. The aluminum flywheel is a metal-sprayed unit forged from the finest aluminum alloys and treated to withstand normal warpage and distortion under the most severe competition condi-

'tions. And it weighs just 10 pounds, meaning increased revability. The Super Rev Lok pressure plate was chosen because of a condition which is common to most factory diaphragm-style assemblies. These stock units tend to hang up on redline power shifts preventing a gear change and causing all kinds of havoc—

Schiefer Super Rev Lok plate is a must for max power shifts.

Matching aluminum flywheel aids revability, is blow-up proof.

Schiefer disc, right, is far beefier than 427 factory unit.

427 CAMARO

especially during a trophy run, Schiefer's improved version boasts constant uniform pressure, increased torque holding capacity with no increase in pedal pressure, no chatter or linkage deflection, and smooth shift control at redline revs. It's the accepted hot setup on the street and strip. The Schiefer disc was chosen because of its durability under competition conditions.

In keeping with the current trend to showy competition-style goodies on the street, Bruce had the Motion Performance crew put mags all around, with Goodyear wide tracks on the front and M&H street slicks on the rear. The

front was also jacked a little to encourage weight transfer and help turn a few heads as well. Rounding off the exterior dress-up mods are NASCAR-type locking hood pins and a set of SS-427 status emblems.

The interior, which is the special extra-cost option, was fitted with a wood steering wheel, a Sun tach and a full array of full-size Sun gauges. The gauges were not installed when the photos were taken, however.

What has Bruce Levinson accomplished by going for a super specialty car instead of a street-type hot rod or an ex-Super/Stock? Well, for one, he owns a respectable transportation piece, tame enough for his wife to

drive to the supermarket. It's also potent enough to go to 60 mph in the 55-second bracket and through the quarter mile in the mid-12's. It can be driven anywhere. It can be serviced by a Chevy dealer or a speed shop. And it carries a modified factory warranty. Bruce Levinson also owns a valuable set of wheels. It can be sold at any time without taking a 50-percent beating and can be simply converted to 327 or 350-cube operation if he wants to sell it as a stocker.

We think this is the only route to travel if you want a genuine dual purpose machine—especially if you have a family. And to boot, Bruce reports that he's getting 13 mpg on the street.

427 CAMARO
(continued from page 20)

thing else on the engine is left completely stock. Rather than go to dual points, Harrell prefers to modify the standard single point 427 distributor by first brazing the vacuum advance linkage to the breaker plate, making it inoperable, and removing any wobble. The standard breaker points are replaced with heavy duty points (No. 1966294) which have a .020-inch thick spring compared to the regular .015-inch spring. Although both are stamped with the same part number the difference can be felt by depressing the breaker points against their spring tension. Harrell uses distributor weight springs designed for six cylinder distributors in conjunction with the regular distributor weights for the 390. He has found this combination will fire accurately up to 7500 rpm. All that is needed to pick up an additional 500 rpm is a piece of foam rubber weather stripping wedged between the floating point and its connector. Harrell then adjusts the advance curve for the type of driving for which the car will be used — drags or street.

For an engine break-in oil, Harrell suggests 20 or 30 SAE Valvoline racing oil combined with one quart of Bardahl. Contrary to popular belief, he has found that the 427's are set up with pretty close tolerances and should accidental overheating occur, there is a good chance of the cylinder walls and pistons getting scuffed. Bardahl has been found to help prevent any scuffing so it's a must for Harrell's super engine.

With all that happiness under the hood, naturally we could hardly wait to give the 427 Camaro a try. On cold mornings it started right up with the automatic choked Holly functioning perfectly. After a short warm up (outside temperature was near 30 degrees during our stay in Chicago), we blasted off down the street. Let me tell you, troops, *that* was an experience!

Almost no foot pressure was necessary to literally paste you to the back of the seat and the engine worked smooth and effortlessly in moving the Camaro through the gears. It's hard to beat the smooth potency of the large displacement stock engine. The manual steering felt completely unaffected by the added 90 pounds of the 427 and had the quick response of a race car, really unbelievable when compared to the production line Camaro.

While the steering was quick and responsive, the suspension seemed just the opposite, giving quick hard jabs as we bounced down a very rough Chicago side street. Unless you live in an area where super-smooth streets are standard equipment, you might find the heavy duty suspension more "heavy duty" than you bargained for. If it's race car suspension you like, then buddy, this is it.

In the corners you had that glued-in feeling, with the Camaro going into them flat and stable with no sway. A little gas and it pulled right through, handling every bit like a racing car rather than a street sedan. There's no doubt about it, that heavy duty suspension really holds the car. The 427 Camaro as a gymkhana machine should be pretty wild.

Not only do the traction bars tie down the rearend on acceleration, but they completely eliminate all bounce when you make a panic stop. The Mono-plate single leaf springs lets the axle housing bounce to a point where it becomes difficult to keep the nose aimed straight. The combination of the metallic brakes and traction bars bring the 427 to an immediate stop in a straight line.

At the strip it was everything you could ask for. Smooth, hard grabbing starts and wailin' runs like an arrow. Several combinations were tried at the Great Lakes Dragway at Union Grove, Illinois, where, in street trim and with muffled exhaust but with the addition of eight-inch M&H slicks on 15x5 rims and 4.56 gears, the Camaro turned in a 11.90 e.t. at 114 mph. This was with the four-barrel Holly carburetor and close ratio four-speed transmission. Changing the single Holly for two Quadrajet carburetors boosted the time to 126 mph in 11.4 seconds.

The potential of this combination seems almost unlimited, leaving a lot of room for experimentation. In street trim the Camaro is a wild, going concern and on the strip it will keep you right in the thick of the competition.

To give you, the customer, even better and faster service, Nickey has recently established Bill Thomas Race Cars in Anaheim, California, as their associate on the West Coast. Thomas will carry the complete Nickey high performance parts line, as well as build 427 Camaros for Western enthusiasts. Like the 427's built in Chicago, the Thomas built cars will carry the Nickey nameplate and will be classified as stock by AHRA. @

Nickey/Thomas Camaro 427

PHOTOGRAPHY: RICK McBRIDE

Any kind of domestic or foreign sedan can be transformed into a successful drag racing machine. The formula is simple enough: lighten the body and chassis as much as is allowable under whatever rules the concoction is to race (or at least whatever you think you can get away with), and shoehorn in the biggest, strongest engine you can lay your hands on. It all has to do with improving the power-to-weight ratio—the more of the former, and the less of the latter, the better the outcome will be.

But can a strong, strip-proven drag car be made into a usable street-and-highway vechicle? Can the transformation process be reversed or modified sufficiently to let the wife run groceries between weekends at the dragstrip? Well, if you have an understanding wife or an enthusiastic girlfriend, the answer is a qualified yes.

If she's willing to put up with the idiosyncrasies of a manually shifted, highly tuned engine, the vagaries of multiple carburetion, and the exertion demanded by super-sticky tires

and no power steering, she can go along with the game and probably make quicker trips to the supermarket than any housewife in the neighborhood. If she's not willing, buy a trailer or a towbar and drag your fun-car to the track with her all-power sedan. The dual-purpose type of car demands patience, understanding, and a willingness to be the center of attention.

A normal 350- or 396- cu. in. Camaro is a reasonably stout performer—although no terror—at the dragstrip. But transfuse this relative

Can a strong, strip-proven drag racing car find happiness as a usable street and highway performer? With a little patience and understanding, you bet your four-barrels.

lightweight with 427 cu. in. of Corvette power and it becomes an entirely new package. Then give that 427 raceworthy preparation and this mutant fairly yelps for straightaway sessions. On paper, and on the dragstrip, it seemed a tremendously potent combination of elements, but could it work, too, under normal street and freeway conditions? That's what we aimed to find out.

The Nickey 427 Camaro is, no doubt about it, an attention-getter. It looks exactly like what it is—a Camaro with Mr. America musculature. It is a very good attempt at making a dual-purpose vehicle for the Sunday sportsman, though all such creations must, of course, ultimately be a compromise. It is, for openers, a Chevrolet Camaro refitted with a 427 cu. in., 425-hp Corvette engine. It carries the nameplate of Chicago's Nickey Chevrolet, although development and most production is by Bill Thomas/Race Cars of Anaheim, Calif.

Our particular Nickey 427 Camaro had had additional treatment in the Thomas shop to make it both a greater attraction and an even better performer. Huge fat black tires on mag-type wheels lent authority to the red coupe's shapely exterior. A specially prepared engine made promising noises from within. The combination of sight and sound was enough to send a drive-in Don Juan right out of his winkle pickers.

When Thomas' Camaro project manager, Pierce Marshall, started up this car it issued that throbbing "rrrrummpppp, rrumpp, rump," ululation characteristic of hot-cam engines. Counterpointing that was the nervous "hummm, grackle, zzzz, snap, gritch, hummm," cacophony peculiar to race-ready engines idling at 1800 rpm. The engine, it seemed, was trying to escape its metallic restraints by sheer pressure of internal combustion.

"Just keep it above 2000 rpm and the plugs will stay clean," Marshall warned as we climbed into this very *(Text continued on page 41; Specifications overleaf)*

The 427 gets a special cam, two 4-barrel carbs, 12.5:1 pistons and tubular headers.

Camaro SS 350 body is used for its heavy-duty underpinnings and Muncie 4-speed.

39

NICKEY/THOMAS CAMARO 427

Manufacturer: Nickey Chevrolet
4501 Irving Park Rd.
Chicago, Illinois 60641

Bill Thomas Race Cars
503 E. Julianna St.
Anaheim, California 92805

Number of dealers in U.S.: 100

Vehicle type: Front-engine, rear-wheel-drive, 4-passenger sports sedan, all-steel integral body/chassis

Price as tested: $5922.00
(Manufacturer's suggested retail price, including all options listed below, Federal excise tax, dealer preparation and delivery charges; does not include state and local taxes, license or freight charges)

Options on test car: Blueprinted Stage III engine ($1500.00), Nickey Camaro wheels ($126.00), Thomas traction bars ($45.00), tachometer ($43.00), competition shock absorbers ($40.00), competition clutch and flywheel ($130.00), explosion-proof bell housing ($68.00), interior group ($5.30), radio ($57.40), quick steering option ($16.00)

ENGINE
Type: Water-cooled V-8, cast iron block and heads, 5 main bearings
Bore x stroke..4.25 x 3.76 in, 107.9 x 95.4 mm
Displacement..............427 cu in, 6994 cc
Compression ratio................12.5 to one
Carburetion........2 x 4-bbl Carter AFB
Valve gear........Pushrod-operated overhead valves, mechanical lifters
Power (SAE).........550 bhp @ 6500 rpm
Torque (SAE).........440 lbs/ft @ 5500 rpm
Specific power output........1.29 bhp/cu in, 78.7 bhp/liter
Max. recommended engine speed...7000 rpm

DRIVE TRAIN
Transmission.....4-speed manual, all-synchro
Clutch diameter.......................11.0 in
Final drive ratio...................3.73 to one

Gear	Ratio	Mph/1000 rpm	Max. test speed
I	2.52	8.6	56 mph (6500 rpm)
II	1.88	11.5	75 mph (6500 rpm)
III	1.46	14.7	96 mph (6500 rpm)
IV	1.00	21.7	141 mph (6500 rpm)

DIMENSIONS AND CAPACITIES
Wheelbase.............................108.1 in
Track...............F: 59.0 in, R: 58.9 in
Length................................184.6 in
Width...................................72.5 in
Height...................................51.0 in
Ground clearance........................5.0 in
Curb weight..........................3340 lbs
Test weight..........................3680 lbs
Weight distribution, F/R............58/42%
Lbs/bhp (test weight)....................6.7
Battery capacity..........12 volts, 45 amp/hr
Alternator capacity..................444 watts
Fuel capacity.........................18.5 gal
Oil capacity............................5.0 qts
Water capacity........................18.0 qts

SUSPENSION
F: Ind., unequal-length wishbones, coil springs, anti-sway bar
R: Rigid axle, monoplate springs with traction arms

STEERING
Type.....................Recirculating ball
Turns lock-to-lock........................4.25
Turning circle.........................43.1 ft

BRAKES
F..9.5 x 2.5-in cast iron drums, metallic linings
R..9.5 x 2.0-in cast iron drums, metallic linings
Swept area.........................268.6 sq in

WHEELS AND TIRES
Wheel size and type............6.0JK x 14-in, chromed mag-type, pressed steel rims
Tire make, size and type.....Casler Scrambler (8.25-14 front, 8.55-14 rear)
Test inflation pressures...F: 32 psi, R: 26 psi

PERFORMANCE
Zero to	Seconds
30 mph	2.1
40 mph	3.0
50 mph	4.0
60 mph	5.6
70 mph	6.9
80 mph	8.7
90 mph	10.2
100 mph	12.4

Standing ¼-mile..........13.9 sec @ 108 mph
80-0 mph panic stop..........263 ft (0.81 G)
Fuel mileage.......7-10 mpg on premium fuel
Cruising range...................130-185 mi

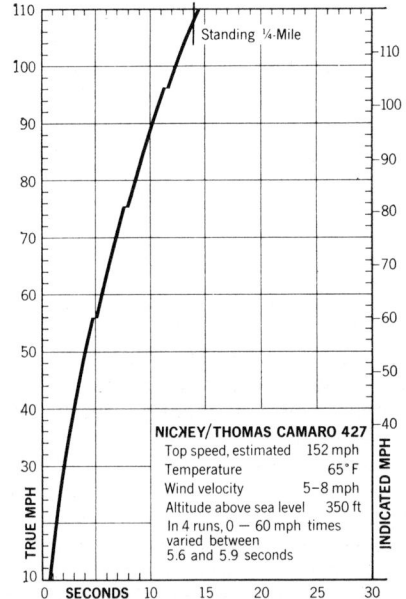

Standing ¼-Mile

NICKEY/THOMAS CAMARO 427
Top speed, estimated 152 mph
Temperature 65°F
Wind velocity 5-8 mph
Altitude above sea level 350 ft
In 4 runs, 0 — 60 mph times varied between 5.6 and 5.9 seconds

TRUE MPH / INDICATED MPH / SECONDS

CHECK LIST

ENGINE
Starting.................................Excellent
Response................................Excellent
Vibration...................................Fair
Noise......................................Poor

DRIVE TRAIN
Shift linkage.........................Very Good
Synchro action.........................Excellent
Clutch smoothness........................Good
Drive train noise........................Good

STEERING
Effort.....................................Fair
Response...................................Fair
Road feel..................................Fair
Kickback...................................Fair

SUSPENSION
Ride comfort...............................Poor
Roll resistance........................Very Good
Pitch control..............................Good
Harshness control..........................Poor

HANDLING
Directional control........................Good
Predictability.............................Good
Evasive maneuverability....................Fair
Resistance to sidewinds................Very Good

BRAKES
Pedal pressure.............................Fair
Response...................................Fair
Fade resistance........................Very Good
Directional stability..................Very Good

CONTROLS
Wheel position.............................Good
Pedal position.............................Fair
Gearshift position.....................Very Good
Relationship...............................Fair
Small controls.............................Good

INTERIOR
Ease of entry/exit.........................Fair
Noise level (cruising).....................Poor
Front seating comfort......................Fair
Front leg room.............................Fair
Front head room............................Fair
Front hip/shoulder room....................Good
Rear seating comfort.......................Poor
Rear leg room..............................Poor
Rear head room.............................Poor
Rear hip/shoulder room.....................Fair
Instrument comprehensiveness....Excellent
Instrument legibility......................Fair

VISION
Forward....................................Fair
Front quarter..............................Good
Side...................................Very Good
Rear quarter...............................Poor
Rear.......................................Fair

WEATHER PROTECTION
Heater/defroster...........................Good
Ventilation................................Fair
Weather sealing............................Good

CONSTRUCTION QUALITY
Sheet metal................................Good
Paint......................................Good
Chrome.....................................Good
Upholstery.................................Fair
Padding....................................Good
Hardware...................................Fair

GENERAL
Headlight illumination.................Very Good
Parking and signal lights..................Fair
Wiper effectiveness........................Fair
Service accessibility......................Good
Trunk space................................Poor
Interior storage space.....................Fair
Bumper protection..........................Fair

NICKEY/THOMAS CAMARO 427

(continued from page 39)

special demonstrator. "And use a red-line of 6500," he casually cautioned as we worked the shift lever over and up for first gear. We got it pointed toward the door, found the right gear slot, and applied a gentle inch of pressure to the throttle pedal. Sheesh! Bill Thomas now has twin 7½-in. black stripes leading out his service entrance.

The first test was out and over a freeway/country road/mountain pass route calculated to reveal the schizophrenic tendencies latent in any kind of car. The freeway portion proved dull, dull, dull. The Nickey Camaro just rumbled along at 65 mph, engine smooth but impatient at 3000 rpm. The stiff steering became more nimble and the ride a bit on the harsh side at anything over 20 mph. So much for legal, 65-mph cruising, although even there the car caused young heads in adjacent cars to turn, fingers to point and knowing smiles to spread. The chromed script "Nickey Camaro" on the foreflanks (even without Nickey's characteristic reversed "k") only confirmed their anticipation.

Off the freeway, then, and onto a two-lane, high-crowned country road. An unlimbering of mind and foot, accompanied with suitable movement of the shift lever, and the Nickey Camaro's true personality emerged. We rushed at curves, downshifting and up-revving, and came out tail swung wide in dirt-track style. Despite the good forward traction of those wide tires, the lightly laden tail showed a marked propensity for switching around. The breakaway point proved a little lower than expected, but was easily learned and controlled by more power and a steer in the direction of the slide.

A charge up the twisting grade to the pass revealed some more quirks in the Nickey Camaro personality: it didn't like to lean in the corners. The large supplemental anti-sway bar under the front end, and Thomas' special rear spring traction kit, had eliminated that common Camaro problem. True, an initial understeering tendency resulting from the weight distribution was pretty powerful and had to be countered with plenty of throttle and wheel action for fastest cornering, but the lack of body roll lent confidence to the driver under such conditions.

So the Nickey Camaro—at least our test car—could be an acceptable handler even though its hefty frontward weight bias would handicap it as a road-racer. We gave the fat footprint tires great credit for keeping the car's ends in proper relationship.

Thumping around the curves on the downgrade side of the pass made us appreciate the fade-resistant metallic-lined drum brakes; we could drive well into the bends before standing good and hard on the binders, then we downshifted and thundered out the other side. Tiring of that game, we resorted to a new one—"trickle" driving. That's where you do everything the easy way, just trickling around, making a minimum of gearchanges, throttle movements, and braking. The idea is to see how little effort you can expend. It didn't work well with the Nickey Camaro, especially in suburban traffic. This was a car that had to be driven. The gears always seemed to need changing, the brakes needed using, the steering had to be twisted. Where freeway driving was too effortless, in-town driving became too busy (busy hands aren't *always* happy hands, Mother!).

Having satisfied ourselves that the Nickey Camaro could be driven as a normal, if energetic, sort of sports sedan, we headed off to the dragstrip for more specific measurements. Our best test run was 13.9 secs. with a trap speed of 108 mph. Starts made with 4000 on the tach, an all-on clutch, and a gently feathered throttle produced the best results. Elapsed times, however, were disappointing. A car such as this should turn in the low 13s. The blame must be on the street mufflers and the wide-ratio 4-speed gearbox. Thus equipped, an unusable 7000 rpm or more was needed for upshift points in order to keep the engine operating within its narrow power range. We didn't remove the mufflers because our object was to measure the car's street performance.

As a point of comparison, the same car and engine set up more for drag racing than street use had turned the quarter-mile in 11.35 secs. and 127 mph previous to our test. There were several important differences, of course, the main ones being a 4.88:1 drive in place of our 3.73 "street" axle, and a close-ratio (2.20 first) transmission. Mufflers were removed from the headers and Casler 9-in. slicks were used on the rear wheels. That kind of dragstrip performance is impressive with a stock-bodied sedan.

Braking tests were next and the metallic-lined drums gave so-so results; no fade problems, but not very fast stopping, either. The extra weight in front obviously unbalanced the brake system, and without a proportioning valve the application of enough pressure to really snub the front wheels caused the rears to lock. Overall effectiveness was low. Chevrolet's front disc option might well be worthwhile despite its extra cost. (Drag racers don't like discs because they're heavier than drums.)

Then it was on to the freeway again for the return to Thomas' headquarters. A few more criticisms were discovered, but mostly they were applicable to all Camaros—like the bucket seats that don't have sufficient lateral support, a high-placed steering wheel and poor rear-quarter vision. Instrumentation was grossly inadequate for such an expensive engine and automobile. The standard speedometer in the left pod was easy enough to read, as was the big electronic tachometer over the right pod, but there were no temperature or pressure gauges, only warning lights. The fuel gauge was hidden by the tach, as well it should have been, for the gluttonous dual Carter AFBs drank gas at the rate of 6.9 mpg—another compromise that must be accepted with the dual-purpose premise.

Just how do you buy one of these machines? After the customer lays down his $3891 at either Thomas' or Nickey's counter, the process begins. Each new Camaro is custom remade to the buyer's own specification, and $3900 usually is only the beginning. Nickey/Thomas has a kit or component for every need, and the wise salesmen stand ready to counsel the customer on what he really should have to go with that fine big engine.

Nickey and Thomas have worked out an ideal coalition for the creation and marketing of 427 Camaros. Thomas does the development work and builds the West Coast-sold cars. Nickey provides the capital and sells in the midwest and east. Both names are well known to racing and performance fans, and both have always been associated with very successful Chevrolet-based equipment. Thomas, as a result, has developed an exceptional line of high performance equipment for Chevrolet cars and engines. He was at work on Camaro components even before the car had made its first public appearance last fall.

Although it's an easy, almost bolt-to-bolt engine swap, certain fillips developed by Thomas make the whole thing work out better. For one, he orders only the SS 350 models because they're delivered with heavy-duty engine mounts, suspensions, axles and the 4-speed Muncie transmission. Additionally, the cars are ordered with the air conditioning system's radiator, fan and shroud for their greater cooling capacity, and with metallic-shoe drum brakes. All these then become standard equipment for the Nickey 427 Camaro. There are certain pecuniary advantages to the remanufacturer. The 350-cu. in. powerplant is sold off the shelf, complete, as a new unit, for something around $500. The HD equipment is much less expensive as original equipment than as add-on-later parts. The metallic-lining brakes are specified because they're about half the cost of the optional disc/drum system ($79 extra).

Bill Thomas, however, has never been one to let well enough alone. If an engine is good—and Chevy's porcupine-head 427 is very, *very* good—Thomas believes in improving it. "We just make it run a little better," he says. He certainly does. His extensive shop facility includes an engine dynamometer, and it delights Bill to see how much "improvement" he can register on it. Would you believe a reliable 550 horsepower? Nothing to it (with open headers). Our test car had one of Thomas' "better running" engines and it was impressive beyond the bare statistics.

The engine modification technique of blueprinting provided the basic improvement; a special Thomas camshaft, a pair of 4-barrel Carter carburetors, 12.5:1 pistons and a set of Thomas' tubular headers brought out the remainder. There was nothing magical about it, just good shopwork. Blueprinting means that engine tolerances are reduced from production-line variation back to design standard, and all reciprocating components are balanced. It is generally considered to yield as much as a 10% improvement over the original horsepower, although just as valuable is its ability to make an engine survive maximum-output.

The Thomas-developed Camaro headers retail in a kit at $139.95. They have equal-length headpipes of 36 inches, collecting into a 2½-inch single outlet for each bank. The camshaft, another Thomas kit, is $164.55 with dual springs, aluminum retainers, special lifters and chrome-moly pushrods. The cam has 310° duration with 84° of overlap and .565 in. lift. Because it was designed to be a drag racing cam, its strength is mostly at the top end. The dual quads on the aluminum high-rise manifold are also top-end-

only equipment, and go for $150.95. All that throttle opening and super-rich jetting are best used for maximum power development rather than flexibility.

If all this seems aimed at the creation of a super-potent drag racing Camaro, that was just what originally was intended. Thomas wanted the hottest kind of demonstrator he could build. So this one carried virtually every component in his catalog. That it was successful as a drag racer was borne out by that 11.35-second timing.

Then came our question: can it be a successful dual-purpose car, too? Back to the shop for detuning. Actually, this process was quite simple. Stock Chevrolet mufflers and tailpipes were clamped on the header extensions. Slightly warmer spark plugs were substituted. The drive gears were changed to 3.73:1, and the wide-ratio Corvette transmission (2.52 first, 1.88 second, 1.46 third) was installed, but only because the close-ratio box had broken. A hefty 1-in. front anti-roll bar (a $29.95 kit) was substituted for the .6875-in. standard bar. A set of Casler "Scrambler" tires were fitted, 8.25-14s in front, 8.55-14s in back.

The Nickey/Thomas Traction Kit ($39.95) was installed, too. This kit consists of two arms which substitute for the bottom spring plates and extend forward to clamp on the spring leaves just short of their front eyes. These act both as radius arms and anti wind-up levers. They do a satisfactory job of taming the stock Camaro's biggest headache, axle hop, but also apparently stiffen the springs to a point just short of rigidity. Forward traction, even at the dragstrip, was outstanding.

With kits and plenty of boodle, then, the Nickey 427 Camaro can be a very acceptable dual-purpose car. It all becomes a matter of what kind of compromise the owner will accept. The stock 427 will produce almost the same kind of performance recorded here, but with greater flexibility at the lower end of the operational scale. However, if he wants that top-end blast, he can have it, too. If outstanding handling would be a requirement, then Thomas could develop yet another kit, based on the Corvette's independent rear suspension, although as yet there's been no demand. The Camaro provides an admirable, attractive base upon which to build a very special type of automobile, and Nickey Chevrolet and Bill Thomas/Race Cars are just the people who can do it. **C/D**

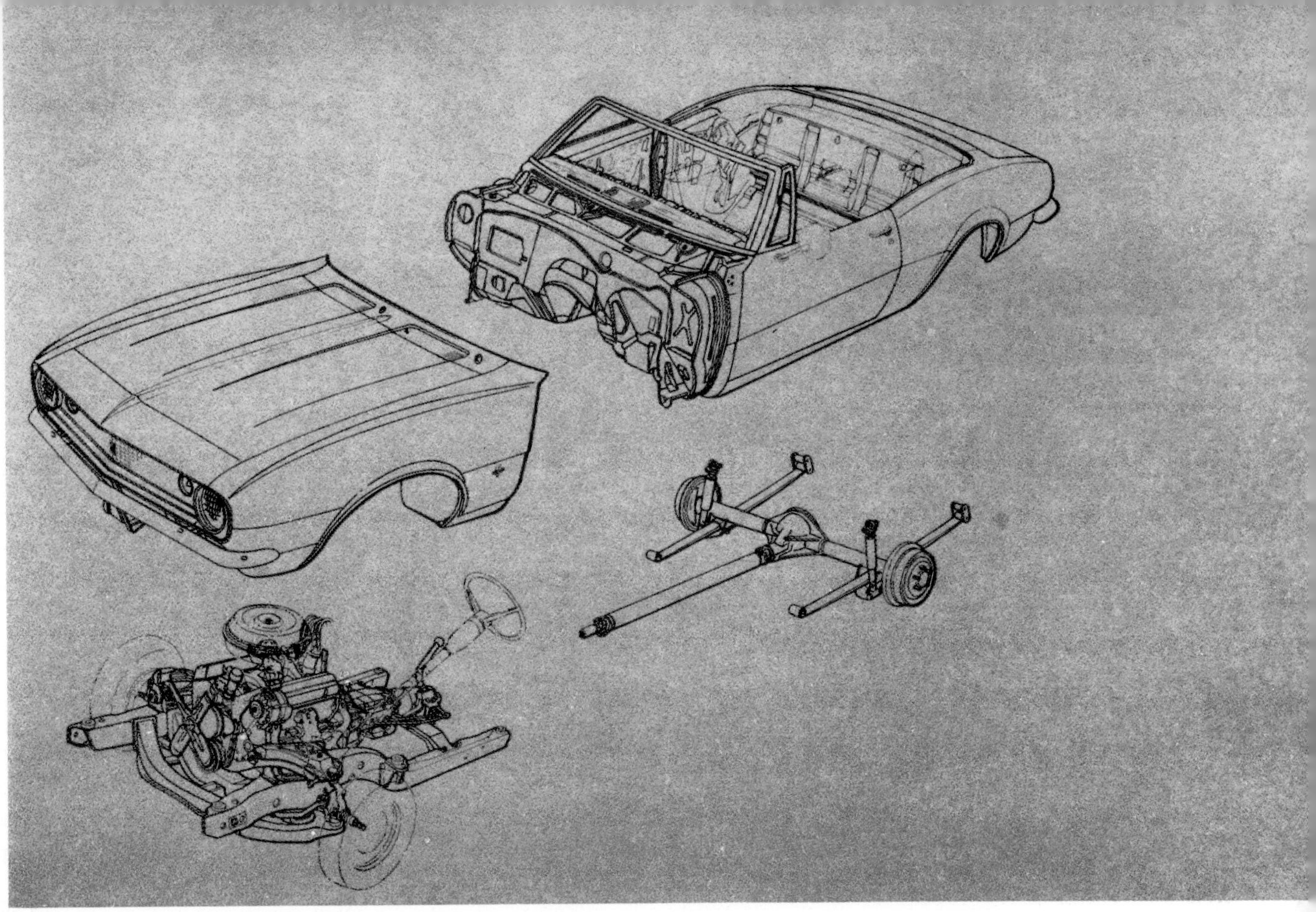

MAJOR COMPONENTS of the 1967 Camaro can be seen in this exploded view. For 1968 it has multileaf rear springs and rearranged rear shock absorbers. The Camaro/Firebird chassis is one of the most computerized engineering designs to date. Below, an engineer with a bank of computers.

ENGINEERING THE
CAMARO

THE FIRST F-CAR prototypes looked like slenderized Chevy IIs. As it's turned out, Camaro and Chevy II still have a lot in common. The Camaro's unitized passenger compartment bolts via four rubber-bushed mounts to a separate stub frame out front. This stub frame carries the engine, transmission, steering, and front suspension, while the body structure supports the rear axle on single-leaf "monoplate" rear springs in cars equipped with lower bhp engines.

Chevrolet engineers picked this half-unitized, half-frame construction for very good reasons. Conventional unit body/chassis structures tend to transmit a good deal of noise and vibration into the driving compartment. Yet the older style separate-body-mounted-on-

CAMARO ENGINEERING

a-full-frame system has the disadvantage of ponderous weight and bulkiness. So by copying the half-and-half plan used so successfully in the Chevy II, the Camaro's power unit could have its own isolated frame while passengers ride in a lightweight, rattle-free cockpit behind it.

Carefully planned curves in the hood and front fenders contribute to front-end stiffness, just as exterior panels from the firewall back add strength to the basic body structure. The Camaro convertible, without a

steel roof to give it torsional rigidity, compensates with double stringers inside the rocker panels but is necessarily weaker than the coupe.

Due to the convertible's normal lack of torsional rigidity, Chevy engineers had to install shake dampers at all four corners. Without these, the structure would cause body shake that eventually might weaken the metal. These dampers consist of oil-filled cylinders with spring-suspended weights inside. When cyclic oscillations begin, the dampers react against this motion, thus smoothing out harshnesses.

In all, the Camaro's half-and-half construction adds up to an extremely tight, rugged, strong, yet lightweight package, capable of accepting any number of different powerplants. And

the Camaro offers more different-displacement engines than any of its rivals: six in all.

Three of these, the 302-, 327-, and 350-cid V-8s, are direct descendants of the highly successful and much favored Chevy 265/283, initially introduced in 1955. Here's how they relate:

Displ.	Bore & stroke
265	3.750 x 3.000
283	3.875 x 3.000
302	4.000 x 3.000
327	4.000 x 3.250
350	4.000 x 3.480

In other words, the 283 is a rebored but also recored 265; the 327 adds a little more bore yet plus 0.25-in. more stroke. For the 350, they stroked the 327 almost another quarter inch. And

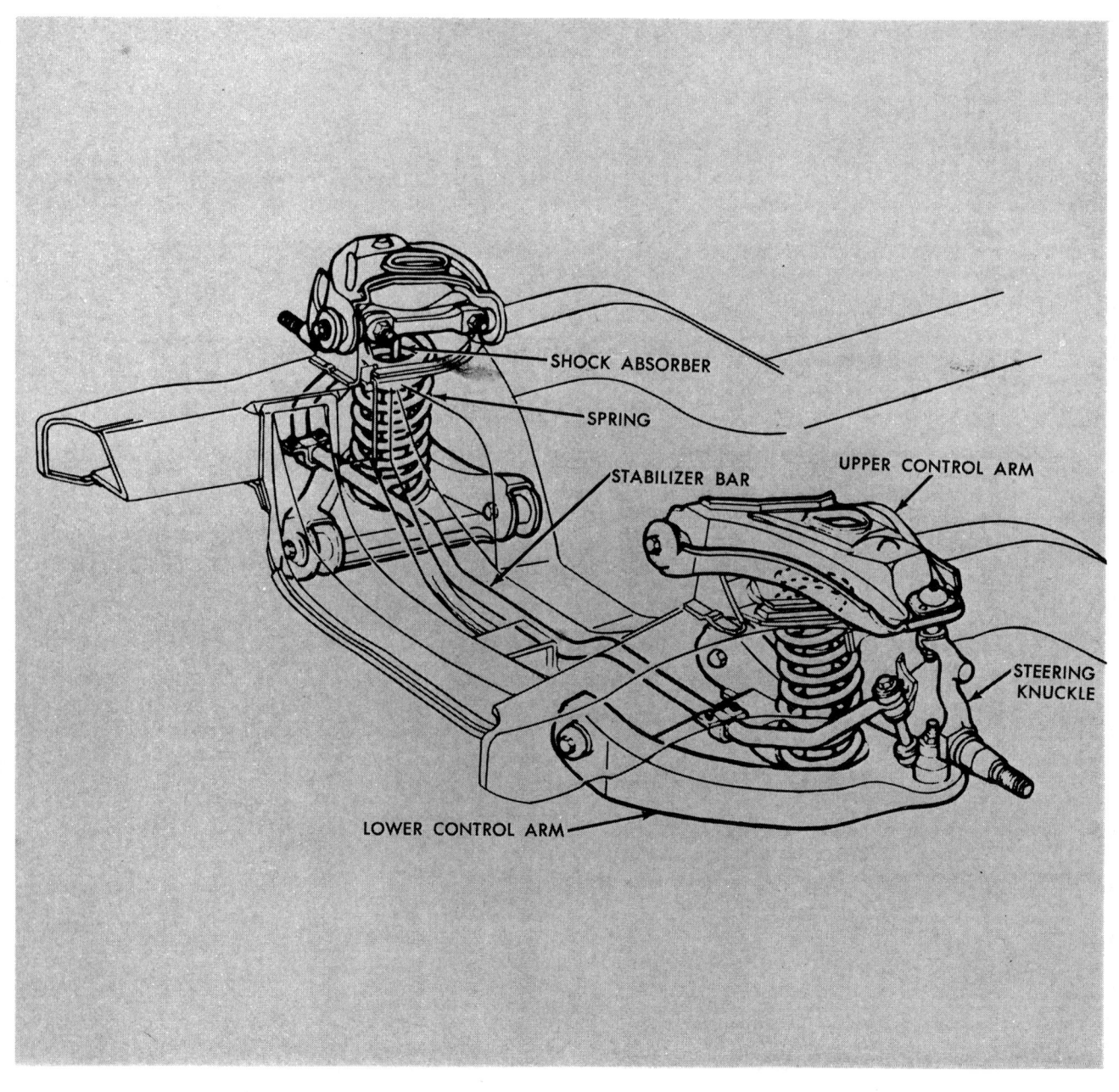

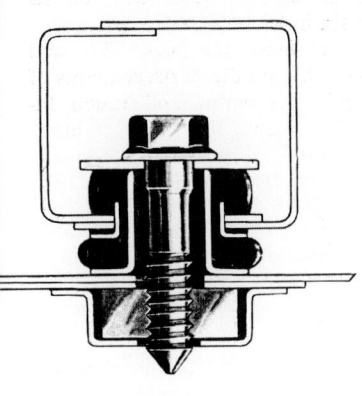

FRONT STUB FRAME is attached to body by four double-biscuit rubber mounts.

finally, to bring about the 302-cid V-8, Chevy engineers merely took the 327 and destroked it by substituting the 283's crank (drop-forged steel, though, instead of the 283's cast-iron one). All in all, these permutations and combinations of the old 283 add up to pretty clever planning. Before you Chevrolet owners run out though, and try to rebore your 283s to make them 327s or 350s, remember that all these engines use entirely different block castings. While outside dimensions are essentially the same, the 283, 302, 327, and 350 use different castings. Same goes for trying to stroke these blocks to make them bigger. The smaller versions don't have enough heft to take the added stroke, even though main-bearing diameters might be the same.

The three smaller Camaro V-8s—302 to 350 cid—all weigh just about the same, around 550 lb. without fly-wheel and clutch. All have 21.75-in. block lengths, 4.4-in. bore centers, a 9-in. face up the side of the block, and most of their parts are interchangeable.

Chevrolet engineers agreed from the beginning that the 283 V-8 had too little low-end torque to please prospective Camaro owners, so they decided not to offer it at all in this series. Instead, the base V-8 became the 327, which now delivers 210 bhp in its economy form (single 1-bbl. carb) or 275 bhp with the 4-bbl.

Most interesting among these engines is the 302-cid version, putting out a conservatively rated 290 bhp at 5800 rpm and easily capable of turning 7000

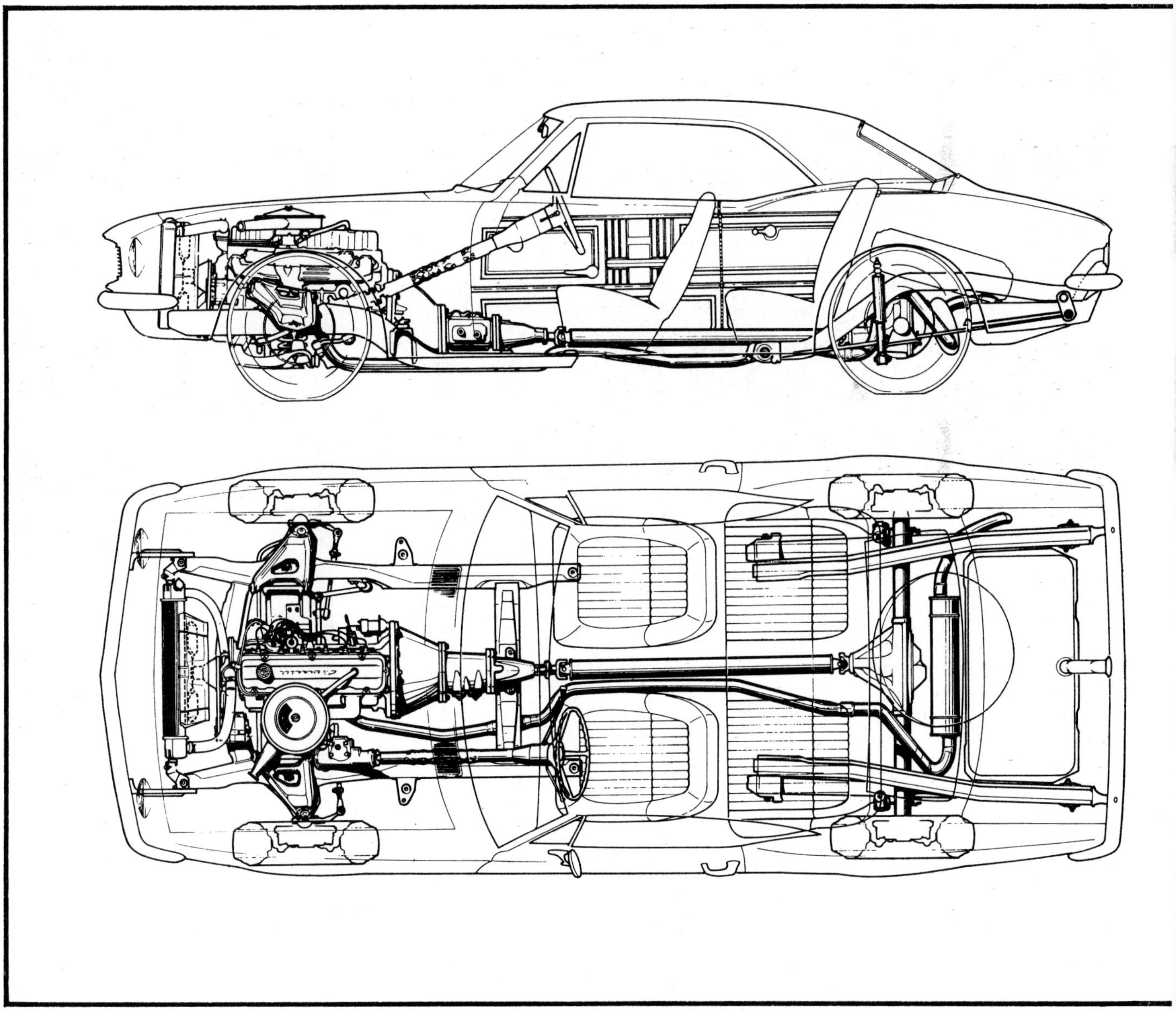

COMPONENT LAYOUT of the Camaro (1967 six cylinder model is shown) shows how tightly things have to be arranged when typically large U.S. power components are installed in a smaller than usual chassis/body structure. Space utilization is very good except for too-small luggage compartment. This definitely needs improving.

CAMARO
ENGINEERING

rpm. The reason Chevy decided to add the 302 to its lineup was so they'd have something to enter in sedan racing class under Sports Car Club of America (SCCA) rules. Maximum Group II displacement under SCCA limits is 305 cid.

Also, the 302 and 350 engines have larger dia mains than the others, rather important. The 302 crank is forged too.

This sedan-racing-intended V-8 brings out the best in Chevrolet's engine-building art. Big-valve heads with 2.02-in. intake valves and 1.60-in. exhausts (0.447-in. lift) make breathing through a Holley Hi-Performance carburetor easy. This carb stands atop a tuned-runner aluminum intake manifold; then gently curved exhaust headers help remove the spent gases.

Huskiest factory-supplied engine, though, is the semi-hemi 396-incher. This powerplant, replacing the larger and heavier W-series powerplants based on the 348/409-cid jobs begun in 1958, delivers 325 bhp in the Camaro and has produced as much as 375 horses in earlier SS Chevelles.

Since Chevy's planners suspected some dealers wouldn't be satisfied even with the 396 and would probably stuff the Corvette 427 into the Camaro, they built enough beef into the body/

chassis structure to take that powerplant, too. In fact, this amazing little car will probably accept even more additional power before structural strengthening becomes necessary.

At the lower end of the engine spectrum, two 6-cylinder units lend genuine economy. Both the 230- and 250-cid powerplants have seven main bearings, and, in addition to their miserliness, are extremely rugged and easy to service. While their outputs aren't spectacular—140 and 155 bhp —they have enough snap to accept a fair number of power accessories and still do a respectable job of making the car perform.

In designing the Camaro's suspension, Chevy engineers called on computers to help. Front suspension was based largely on Chevelle experience, and the two systems look very much the same. It's a short-and-long-arm arrangement with coil springs sandwiched in between. Front shocks rest inside these coils.

More interesting, though, is the rear springing setup. As mentioned, the engineers initially decided on single-leaf springs for the Camaro. In all similar monoplate springs put under Chevy IIs, not a single case of breakage has ever been reported. Yet while amply strong, these springs caused jitter prob-

lems, axle hop, and tended to wrap up with the higher-horsepower engines. So engineers saved themselves a lot of hit-and-miss experimentation by feeding requirements into computers and letting these figure out the best solutions. As it turned out, revised spring eyes, various traction bars, and differently angled rear shocks took care of all problems for the 1967 model run. In fact, the more upright shock absorbers specified by the computers even let engineers install a larger-than-planned gas tank, giving the Camaro an improved cruising range.

For 1968, though, Chevy decided to make multi-leaf rear springs standard on Camaros with 327-cid V-8s and 4-speed transmissions, and on those with 350- and 396-cid engines. All the rest still use monoplate rear leaves, same as before.

Alex C. Mair, Chevy's engineering director, summed up the suspension improvements this way: "Handling actually came out better than we'd intended. The car's directional stability, particularly, is better than predicted. And there are several reasons for this. One is that we widened the Camaro's tread during its late development stages —set the rear wheels farther apart. That caused some gain in handling capabilities."

In the area of braking, big 11-in. discs offer the greatest stopping power when combined with sintered metallic drum brakes on the rear. This setup is standard with the Z-28 (302-cid sedan racing) package and may be ordered optionally on other Camaros. Or you can get the standard 9.5-in. drum system at all four wheels with either metallic or regular linings. The disc system gives 332.4 sq. in. of total swept area, while drums all around give 268.6 sq. in.

Basic Camaro transmission is an all-synchro 3-speed with floorshift. A 2-speed Powerglide automatic and 4-speed manual are optional with all engines except the 302 and the 396. The 302 comes only with a heavy-duty 4-speed, while the 396 can be ordered with the 3-speed Turbo Hydra-Matic.

An extremely wide array of trim, handling, and power packages (see list, page 74) makes the Camaro one of the easiest and, at the same time, hardest cars to tailor. There's almost too much to choose from. But with patience, there's no reason why any new Camaro buyer can't be completely satisfied. ∎

DISC BRAKES are optional on the front of all Camaros, and at front and rear of Z-28 version (which has phenomenal deceleration).

CAMARO Z-28

As near as Chevrolet can come
to being in racing without being in racing

WHO SAYS GM isn't racing? If the Z-28 isn't a bona fide racing car—in street clothing for this test—then we've never seen one. Chapter IV, Touring Cars, Group 2, Appendix J, FIA International Sporting Code requires that 1000 of the Group 2 sedans (more popularly known in this country as Trans-Am sedans) be series produced and that's the reason for the Z-28's being. No question that the Z-28 is doing its stuff either: two of them followed two Porsche Group 6 prototypes home at Sebring to finish third and fourth overall and win the Trans-Am category of Sebring's 12 hours.

Z-28 is a typical Chevrolet code designation for a performance package which adds $400.25 to the basic Camaro 6-cyl coupe price of $2694 and includes a 5-liter (302-cu-in.) V-8, slightly modified spring rates, quicker steering and identification trim. To that $400 is added $100.10 for power-assisted disc front brakes and $184.35 for 4-speed manual transmission. Our test car was equipped with still-quicker (17:1 overall) steering at $15.80, limited-slip differential ($42.15), power steering ($84.30) and a host of such items as interior trim packages, custom steering wheel, deluxe seat belts (!) and a tack-on fiberglass "spoiler" for a total price of $4435. To the basic Z-28 package can be added a whole range of racing parts, available from Chevrolet dealers.

The chrome-trimmed 302 engine is a combination of the 327 block (4.00-in. bore) and the 283 crankshaft design (3.00-in. stroke)—but with a stronger, forged steel crank instead of the cast nodular iron one of the 283 and other "mild" Chevrolet engines. The 302 comes in a rather wild state of tune: its standard camshaft gives 346° duration for both intake and exhaust valves—118° of overlap. Mechanical lifters and 1.50:1 rockers give a valve lift of 0.485 in; an optional

cam provides equivalent overlap with greater lift, but our test car had the standard one. Compression ratio is 11.0:1 and carburetion is by a single Holley 4-barrel rated at 800 cu ft/min. Our test car also had, as its only item from the "dealer" list, a set of tuned headers (installed by Bill Thomas Race Cars, as these are never installed at the factory anyway) which add another $200.

The engine makes no bones about its character. It idles lumpily at 900 rpm and has very little torque below 4000 rpm, considering the car's great weight (3355 lb). It does start easily from cold, however, the automatic choke putting idle speed up to something like 1500 rpm; very little mechanical

SCALE: 10" DIVISIONS

PRICE

Basic list.................$3527
As tested.................$4435

ENGINE

Type....................V-8, ohv
Bore x stroke, mm....101.6 x 76.3
  Equivalent in........4.00 x 3.00
Displacement, cc/cu in..4949/302
Compression ratio..........11.0:1
Bhp @ rpm......est. 350 @ 6200
  Equivalent mph...........115
Torque @ rpm, lb-ft,
  ....est. 320 @ 4200
  Equivalent mph............77
Carburetion.one Holley 1.69-in. 4V
Type fuel required.......premium

DRIVE TRAIN

Clutch diameter, in........10.3
Gear ratios: 4th (1.00).....4.10:1
  3rd (1.27).............5.21:1
  2nd (1.64).............6.72:1
  1st (2.20).............9.02:1
Synchromesh..............on all 4
Final drive ratio..........4.10:1
Optional
  ratios.....3.55, 3.73, 4.56, 4.88:1

CHASSIS & BODY

Body/frame: unit steel construction
Brake type: 11.0-in. vented disc
  front; 9.0 x 2.0-in. drum rear
  Swept area, sq in.........332
Wheels.......steel disc, 15 x 6JK
Tires..Goodyear WideTread E70-15
Steering type....recirculating ball
  Overall ratio...........17.0:1
  Turns, lock-to-lock........2.8
  Turning circle, ft.........37.0
Front suspension: unequal-length
  A-arms, coil springs, tube
  shocks, anti-roll bar
Rear suspension: live axle on
  multileaf springs, tube shocks

OPTIONAL EQUIPMENT

Included in "as tested" price:
power steering, fast steering
ratio, limited-slip, headers, extra
instruments, AM radio, tinted
glass, various trim items.
Other: many performance, braking, handling items from dealers.

ACCOMMODATION

Seating capacity, persons.......4
Seat width,
  front/rear.......2 x 21.0/54.5
Head room, front/rear...36.3/34.0
Seat back adjustment, deg.....0
Driver comfort rating (scale of 100):
  Driver 69 in. tall...........95
  Driver 72 in. tall...........80
  Driver 75 in. tall...........60

INSTRUMENTATION

Instruments: 120-mph speedometer, 7000-rpm tachometer, ammeter, water temp, fuel level,
oil pressure.
Warning lights: directional signals,
high beam, low fuel, parking
brake.

MAINTENANCE

Engine oil capacity, qt........5.0
  Change interval, mi.......6000
Filter change interval, mi...12,000
Chassis lube interval, mi.....6000
Tire pressures, psi........24/28

MISCELLANEOUS

Body styles available: coupe only.
Warranty period, mo/mi: 24/24,000
  (60/50,000 on driveline)

GENERAL

Curb weight, lb...........3355
Test weight...............3695
Weight distribution (with
  driver), front/rear, %....56/44
Wheelbase, in.............108.0
Track, front/rear.......59.6/59.5
Overall length.............184.7
  Width.................72.5
  Height................51.4
Frontal area, sq ft.........20.7
Ground clearance, in........5.1
Overhang, front/rear....36.6/40.0
Usable trunk space, cu ft.....8.3
Fuel tank capacity, gal.......18.5

CALCULATED DATA

Lb/hp (test wt).............10.6
Mph/1000 rpm (4th gear)....18.4
Engine revs/mi (60 mph)....3260
Piston travel, ft/mi.........1630
Rpm @ 2500 ft/min.......5000
  Equivalent mph.........93
Cu ft/ton mi...............154
R&T wear index.............53
Brake swept area sq in/ton....180

ROAD TEST RESULTS

ACCELERATION

Time to distance, sec:
  0-100 ft.................4.2
  0-250 ft.................5.9
  0-500 ft.................8.4
  0-750 ft................10.7
  0-1000 ft...............12.8
  0-1320 ft (¼ mi).........14.9
Speed at end of ¼ mi, mph...100
Time to speed, sec:
  0-30 mph................3.4
  0-40 mph................4.3
  0-50 mph................5.5
  0-60 mph................6.9
  0-70 mph................8.8
  0-80 mph...............10.4
  0-100 mph..............15.0
  0-120 mph..............21.8
Passing exposure time, sec:
  To pass car going 50 mph....3.5

FUEL CONSUMPTION

Normal driving, mpg.......9-13
Cruising range, mi.......175-240

SPEEDS IN GEARS

4th gear (7100 rpm), mph.....132
3rd (7500)..................113
2nd (7500)..................85
1st (7500)..................63

BRAKES

Panic stop from 80 mph:
  Deceleration, % g..........59
  Control....................fair
Fade test: percent of increase in
  pedal effort required to maintain
  50%-g deceleration rate in six
  stops from 60 mph.........nil
Parking brake: hold 30% grade.yes
Overall brake rating.........fair

SPEEDOMETER ERROR

30 mph indicated.....actual 31.0
40 mph...................40.6
60 mph...................59.8
80 mph...................79.3
100 mph..................98.6
Odometer, 10.0 mi....actual 10.0

ACCELERATION & COASTING

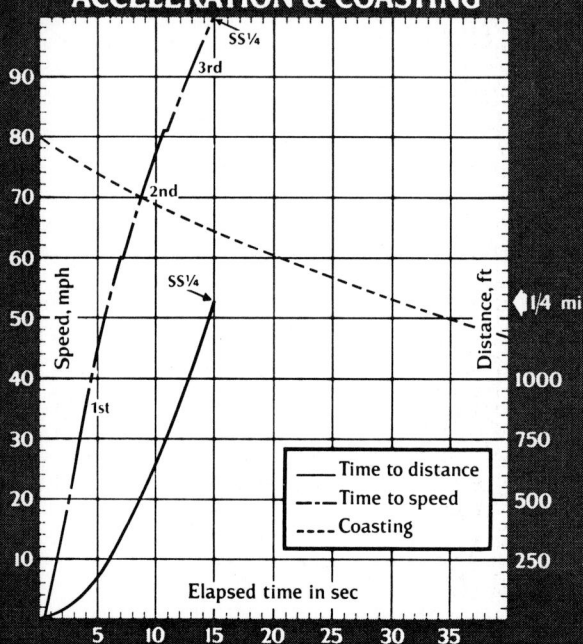

Speed, mph — Distance, ft

SS¼ — 3rd — 2nd — 1st — SS¼ — ¼ mi

—— Time to distance
– – Time to speed
···· Coasting

Elapsed time in sec

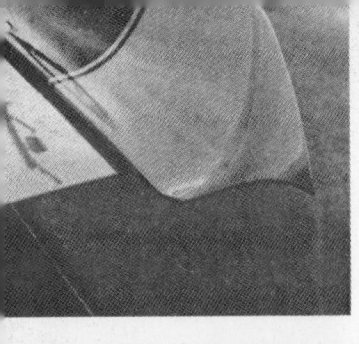

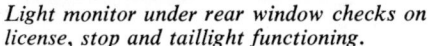

Light monitor under rear window checks on license, stop and taillight functioning.

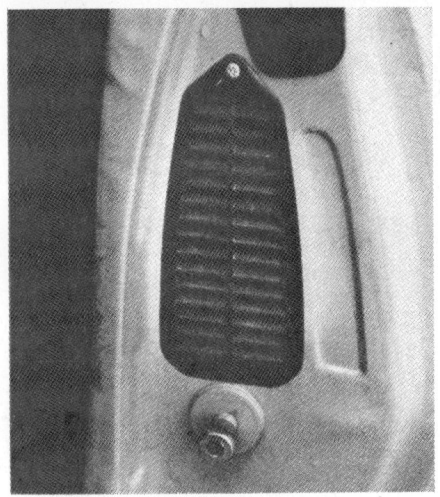

Flow-through ventilation system exits stale air through trunk and then out doorjambs.

noise is heard from under the hood though the headers give it a nice tingly sound. Getting off the line in acceleration tests the car slews smoothly to the right but, even with the clutch dropped at 4500-rpm, it gets off to a relatively leisurely start until the engine can get up over the 4000 hump. Then, hold on! From there it revs so freely that it seems it could go on forever. We found 7500 to be a good shift point as far as the engine was concerned, but above 7000 (63, 85 and 113 mph in 1st, 2nd, 3rd resp.) the diaphragm-spring clutch became reluctant to re-engage after our rather unmerciful shifts—so we were forced to use 7000. Chevrolet rates this engine at 290 bhp @ 5800 rpm, which may be true as far as it goes, but we think the curve keeps climbing to something more like 350 bhp @ 6200. We even think the torque rating of 290 lb-ft @ 4200 is conservative.

The Z-28 can be had with a range of final-drive ratios from 3.55 to 4.88:1. The standard 3.73:1 would seem best for road use, if anyone seriously contemplates using the car for such; our car had a 4.10:1, probably good for road racing but getting the engine into much noise and vibration at freeway speeds. Fuel economy? Who cares?—but 11-mpg average, if you must know.

The 4-speed gearbox is the familiar and beloved Muncie unit with the not-so-beloved Muncie shift linkage. This example wasn't as bad as some we've tried, but the linkage is stiff and notchy, characteristics aggravated by Chevrolet's dumb sliding-plate shift lever seal. Fortunately the latter

comes only with the optional console, a conventional rubber boot doing the job when no console is ordered. Our test car had the optional close-ratio (2.20:1 1st) box, appropriate for the high numerical final drive. An 11.0-in. clutch with 2450-2750-lb spring pressure (larger and stronger than that supplied with the 396 engine) is still sufficiently light for the average male to operate easily, though as mentioned earlier it does have some trouble at high engine revs.

Though the FIA regulations allow the use of any springs and shocks in racing, the road Z-28 sticks surprisingly close to stock Camaro suspension. The front coil springs (112 lb/in. at the wheel) and $^{11}/_{16}$-in. anti-roll bar are left unchanged from everyday Camaro 327s. At the rear spring rates are considerably stiffened: 25% more than for the 396 model, or 131 lb/in. at the wheel. Like the 396, 350 and the 4-speed 327 Camaros (and all Firebirds) the Z-28 uses multi-leaf rear springs instead of the standard Camaro single-leaf ones; this is a necessary move for any engine-transmission combination likely to be delivering great shock loads to the rear axle as the single-leaf jobs don't do much in the way of controlling axle motion about its own horizontal centerline.

With its E70-15 wide tires the Z-28 is a stable, near-neutral car that has no trouble setting excellent lap times around any reasonably smooth course. The trick with a car like this, with all that torque available in the right gear, is to find that point where you're using just enough throttle to get it around a turn neutrally rather than plowing or spinning out. Our

CAMARO Z-28

test car had power steering, which would be a must with all the weight and the 17:1 ratio, but this is so lacking in feel that one has to learn to drive without the help of feedback from the tires . . . and that's not much fun.

Lining material for the Z-28 brakes, otherwise no different from the normal power disc/drum Camaro option, is harder and thus eliminates any trace of fade in our usual fade test as well as putting the pedal efforts up—to the benefit of pedal feel—a bit. Proportioning isn't good for panic stops in the unladen car: we got only 19 ft/sec/sec or 0.59-g by jamming on the brakes at 80 mph. By controlling them carefully we got up to 24 ft/sec/sec. Again, FIA rules allow changes that will make the brakes more satisfactory, and special parts are on the dealer option list.

The stiffer rear springs (which also add frictional harshness over single-leafs) and the tighter shocks that come with the package do give the Z-28 the general sort of "clumpity-clump" ride we've come to expect of Ponycars with handling packages, and the unit body of the Camaro isn't as resistant to rattles and squeaks as its weight would indicate. But the ride isn't super-stiff either and we found it to be gen-

erally acceptable. We can say little about the driving position or comfort that we haven't said over and over about Ponycars; the steering wheel is close, the seats mediocre and not adjustable for back angle, vision to the front excellent (except for a too-high, too-long hood) and to the rear poor; the minor instruments, add-on options, not well placed for reading; excellent heating and ventilation, aided here by air exits in the doorjambs. The separate lap and diagonal seat belts generally adopted by Detroit this year (including this Camaro) are bothersome to use and messy to not use; we presume this will be taken care of in a year or so when the designers get around to putting a little ingenuity to the problem of passenger restraint.

Surprisingly enough, the usual Chevrolet warranty—2 years or 24,000 miles on the car in general and 5 years/ 50,000 on the drivetrain—applies to the Z-28. Servicing is needed only every 6000 miles, and there's enough room around the compact engine that the enthusiastic owner won't be discouraged from a little tuning on his own. The car is straightforward throughout.

The Z-28 offers a lot of performance for the money; how many 4-seat cars can you name that will do the ¼ mile in 14.9 sec, hit 142 mph and cost just $4435? On the other hand, it's not what we'd call "tractable" and, despite its stability and performance, it's pretty clumsy to drive. However, Chevrolet obviously achieved what they set out to do—namely, build a race-winning Trans-Am sedan.

Photos/Randy Holt, Jr.

CAMARO Z-28

We walked into a Chevrolet dealer's showroom recently to inquire about Z-28 literature, and the salesman asked, "What's a Z-28?" Now maybe you have an idea why there aren't many on the streets. However, there's probably a better reason than that. The Z-28 Camaro isn't really designed for street use, although it has many characteristics that make it twice the automobile than any other Camaro.

One of the biggest drawbacks for street driving with the Z-28 is the engine. It takes a lot of revs to get the car moving from a dead stop. Consequently, you end up taking off faster than you really intended. The engine doesn't begin working well until 4000 rpm; then it's really beautiful — but 4000 rpm is well past most speed limits.

The Z-28 was originally made to run the Trans-American championship. A certain number had to be built before the car was eligible for competition. The 1968 version, which is only a package instead of a completely new car, has four-wheel disc brakes, which help immensely in keeping the car stable by improving the overall high-speed braking power.

We were most impressed with the handling qualities: the car tracked well, steering was quicker than normal, and most of the rear axle tramp has been eliminated with the help of multi-leaf springs. From a performance standpoint, the 302-cubic-inch engine performed much better than any street 327 or 350 we ever drove. Horsepower is rated at 290 at 5800 rpm.

As the advertisement says, "It's the closest thing to a 'Vette yet." And that's the way we feel, too. With the basic Z-28 package, the car sells for about $3100 in Detroit.

51

CAMARO Z/28

A pint-size engine with the heart of a tiger gives it a Supercar's performance and sports car's handling

THE Z/28 CAMARO is not cast in the mold of current Detroit space capsules. It is noisy, almost scary in its response to all controls, and delivers a steady barrage of soft blows to the hindsides of its occupants.

The owner seeking insulation from his roadway environment, the man who wants to be transported to his destination with a minimum of effort and conscious involvement will find the Z/28 totally unsatisfactory. This Camaro needs to be driven, in every sense of the word. To the man capable of extracting them, the Z/28 has a storeroom of treasures. For enjoyment-per-dollar, the Z/28 must be one of the bargains of this decade, not because

it's inexpensive (about $3400 base price with Z/28 options) but because it's so excitingly roadable.

The Z/28 Camaro is not a new car. This same basic package formed the basis for the impressively fast sedan racers which conquered the factory-backed Mustangs in the last two races of the 1967 Trans-Am season, and finished 1-2 in the Trans-Am class in the 1968 Sebring 12-hour race. But, in 1967, Chevrolet treated the Z/28 as an illegitimate son. The car has been available for racing for one with the patience and perseverance to wait out delivery of such a rare animal. Yet this fact seems to have escaped many Chevrolet dealers. Some of Chevrolet's

smaller dealers didn't even know such a package existed. For them, Z/28 was a cryptic symbol that wide-eyed enthusiasts murmured to baffled sales personnel when discussing Camaros.

For 1968, Chevrolet has decided to market the Z/28 as a true street machine, a recognized addition to the Camaro model lineup. In line with this marketing decision, some of the more costly and impractical (for production) items from the sedan racing package have been relegated to the status of dealer-installed options. Among these are steel tubing exhaust headers (which we had on our test car), fresh-air pickup from the cowl plenum chamber, and some of the other racing or ultra-performance parts.

The resulting production package is an acceptable everyday transportation vehicle, if the owner is willing to accept a certain amount of low-speed fussiness and lack of torque for easy takeoffs. Above 30 mph, on winding mountain roadways, back country lanes and the rest of the types of roads which make driving worth doing, the Z/28 Camaro is an exhilarating vehicle. Handling is excellent, cornering power is exceptional, and acceleration through the close-ratio gears is amazing for a small engine. There is something very satisfying about a small-displacement engine producing big-displacement power. You feel like the engine is doing something, not lumbering along wasting space and operating inefficiently. The Z/28 engine is a jewel, an outstanding performer by any yardstick. The chassis is even well matched to the powerplant. Suspension, drive train and brakes are all intended to complement the engine, and they do an admirable job.

The Z/28 powerplant has 302 cid, obtained by installing the crankshaft from the old 283-cid V-8 in a current 327-cid cylinder block. Bore and stroke are 4.00 x 3.00 in., giving low piston speed at high revs and a bore large enough to accommodate big

PHOTOS BY SCOTT MALCOLM

CENTER-PIVOT Holley carburetor atop aluminum high-rise intake manifold gives 302-cid Z/28 engine tremendous high-speed power output, up to 7000 rpm.

valves. Cylinder heads used on the Z/28 are the units produced for the 327-cid/350-bhp engine available in Corvettes and Chevy IIs in 1967. Valve sizes are 2.02 in. intake, and 1.60 in.

exhaust. Stiff valve springs are used, along with a mechanical-lifter valve train. A maximum of 7200 rpm was reached during the test period, and even at this speed there wasn't a

FOUR WHEEL drifts or tail-out powerslides can be done in complete control with the roadable Z/28 Camaro. Standard Z/28 suspension and tire package give high cornering power, yet break-away is smooth and predictable.

trace of valve float to be heard.

Two camshafts are cataloged for the Z/28 engine, but only one is mild enough to be considered for street use. The test car had this standard camshaft, which has duration figures of 346°, intake and exhaust. Valve lift is 0.460 in., for both intake and exhaust, at normal operating clearances. The long duration figures made themselves evident in the test car's lumpy 1000-rpm idle, and its lack of torque below 3000 rpm. From 4000 to 7000 rpm, power output was fantastic. Quarter-mile elapsed times were below 15 sec., despite a very slow takeoff. Trap speeds of well over 100 mph indicate the power at the high end. To reach 100 mph in the quarter, with 302 cid, in a car weighing almost 3700 lb. with two-man test crew is almost incredible.

The Z/28 engine's 290-bhp rating must be taken with a grain of salt. Dynamometer tests on similar engines, completely stock but with all clearances optimized, indicate a true power potential of about 400 bhp.

The standard rear axle ratio is

INTERIOR PACKAGE is attractive, and most controls are conveniently placed. Testers disliked Muncie shift linkage and console-mounted instruments.

1968 CAMARO Z/28

DIMENSIONS

Wheelbase, in.	108.0
Track, f/r, in.	59.6/59.5
Overall length, in.	184.6
width	72.3
height	50.9
Front seat hip room, in.	20.5 x 2
shoulder room	56.7
head room	37.0
pedal-seatback, max.	40.0
Rear seat hip room, in.	54.1
shoulder room	53.6
leg room	29.2
head room	36.7
Door opening width, in.	38.2
Trunk liftover height, in.	31.6

PRICES

List, FOB factory	$2694
Equipped as tested	$4086

Options included: Z/28 engine and performance package, 4-speed trans., light monitor system, power disc brakes, steering; am radio, quick steering, console; air spoiler.

CAPACITIES

No. of passengers	5
Luggage space, cu. ft.	8.3
Fuel tank, gal.	18.5
Crankcase, qt.	4
Transmission/dif., pt.	3/3.5
Radiator coolant, qt.	16

CHASSIS/SUSPENSION

Frame type: Unitized, front stub.
Front suspension type: Independent by s.l.a., coil springs, telescopic shock absorbers.

ride rate at wheel, lb./in.	n.a.
antiroll bar dia., in.	0.688

Rear suspension type: Hotchkiss live axle, multileaf springs, staggered shock absorbers.

ride rate at wheel, lb./in.	n.a.

Steering system: Integral assist recirculating ball gear, parallelogram linkage behind front wheels.

overall ratio	17.0:1
turns, lock to lock	2.8
turning circle, ft. curb-curb	37.0
Curb weight, lb.	3355
Test weight	3695

Distribution (driver),

% f/r	55.7/44.3

BRAKES

Type: Disc front, single leading shoe, cast iron drum rear.

Front rotor, dia. x width, in.	11.0 x 2.21
Rear drum, dia. x width	9.5 x 2.0
total swept area, sq. in.	332.4

Power assist: Integral vacuum.

line psi at 100 lb. pedal	790

WHEELS/TIRES

Wheel rim size	15 x 6
optional size	none
bolt no./circle dia. in.	5/4.75

Tires: Goodyear Wide Tread GT.

size	E70-15
normal inflation, psi f/r	24/28
Capacity @ psi	4980 @ 24/28

ENGINE

Type, no. of cyl	ohv 90° V-8
Bore x stroke, in.	4.00 x 3.00
Displacement, cu. in.	302
Compression ratio	11.0:1
Fuel required	premium
Rated bhp @ rpm	290 @ 5800
equivalent mph	107
Rated torque @ rpm	290 @ 4200
equivalent mph	77

Carburetion: 1x4 Holley.

throttle dia., pri./sec.	1.69/1.69

Valve train: Mechanical lifters, push-rods and overhead rocker arms.

cam timing deg.,

int./exh.	60.8-105.3/108.8-57.3
duration, int./exh.	346.2/346.2

Exhaust system: Dual, transverse mount reverse-flow muffler.

pipe dia., exh./tail	2.25/2.25
Normal oil press. @ rpm	40 @ 1500
Electrical supply, V./amp.	12/37
Battery, plates/amp. hr.	54/45

DRIVE TRAIN

Clutch type: Single dry disc, diaphragm-type pressure plate.

dia., in.	11.0

Transmission type: Four-speed fully synchronized.

Gear ratio 4th (1.00:1) overall	4.10:1
3rd (1.27:1)	5.21:1
2nd (1.64:1)	6.72:1
1st (2.20:1)	9.02:1

Shift lever location: Console.
Differential type: Hypoid, limited-slip.

axle ratio	4.10:1

DEEP GEARING is evident in picture of indicated 130 mph speed at 6300 rpm. Great Z/28 engine was still pulling at this speed, and running free and smooth.

SPOILER on deck lid is appropriate addition to race-oriented Z/28.

3.73:1, and although this seems like a very high ratio, it is a good choice for the engine's high-speed characteristics. The test car had a 4.10:1 rear axle, and this was a little too high. It seems strange at first to be driving down the road at 70 mph with about 3500 rpm showing on the tachometer. But the Z/28 engine is still only halfway up in its operating range, instead of being nearly wound out as are most domestic production V-8s. Also, the Z/28 runs at 3500 rpm very freely and effortlessly. The standard 3.73:1 ratio would have eased cruising, without sacrificing too much acceleration, and we would have preferred this ratio in the test car.

The only operational flaw in the test car's power train was a peculiar unbalance which caused a severe cyclic vibration from 3500 rpm up. This may have been a clutch/flywheel assembly unbalance condition. While it was very annoying, the unbalance did not seem to affect performance or the engine's willingness to run to high speeds.

Behind the Z/28's free-revving engine is the familiar close-ratio Muncie gearbox. This is one of the few American automobiles that genuinely needs close ratios. The Z/28 engine has to be kept up in speed, due to the lack of low-speed torque. The close-ratio

CAR LIFE ROAD TEST

ACCELERATION & COASTING

(Graph: MPH vs ELAPSED TIME IN SECONDS, showing 1st, 2nd, 3rd, 4th gears, QUARTER MILE marked at 4th gear around 100 mph)

CALCULATED DATA
Lb./bhp (test weight) 12.7
Cu. ft./ton mile 154.3
Mph/1000 rpm (high gear) 18.4
Engine revs/mile (60 mph).... 3260
Piston travel, ft./mile........ 1630
CAR LIFE wear index 53.2
Frontal area, sq. ft. 20.8

SPEEDOMETER ERROR
30 mph, actual 30.6
40 mph 40.6
50 mph 50.2
60 mph 59.3
70 mph 69.7
80 mph 78.9
90 mph 88.7

MAINTENANCE
Engine oil, miles/days..... 6000/120
oil filter, miles/days..... 6000/120
Chassis lubrication, miles....... 6000
Antismog servicing, type/miles .. replace PCV valve/12,000, tighten belts, clean air injection system, tune check/12,000
Air cleaner, miles. ...replace/24,000
Spark plugs: AC44.
gap, (in.) 0.035
Basic timing, deg./rpm...4BTC/700
max. cent. adv.,
deg./rpm.............. 32/4400
max. vac. adv., deg./in. Hg.15/17
Ignition point gap, in.......... 0.019
cam dwell angle, deg........ 28-32
arm tension, oz............ 19-23
Tappet clearance,
int./exh..........0.025/0.025
Fuel pressure at idle, psi........ 5.0
Radiator cap relief press., psi.... 15

PERFORMANCE
Top speed (7200), mph.......... 133
Test shift points (rpm) @ mph
3rd to 4th (7000) 101
2nd to 3rd (7000)............. 78
1st to 2nd (7000) 58

ACCELERATION
0-30 mph, sec................... 3.5
0-40 mph 4.5
0-50 mph 5.7
0-60 mph 7.4
0-70 mph 8.9
0-80 mph 10.5
0-90 mph 12.5
0-100 mph 14.2
Standing ¼-mile, sec.......... 14.85
speed at end, mph......... 101.4
Passing, 30-70 mph, sec........ 5.4

BRAKING
Max. deceleration rate from 80 mph
ft./sec./sec................... 25
No. of stops from 80 mph (60-sec. intervals) before 20% loss in deceleration rate .... 8-no loss
Control loss? Slight.
Overall brake performance: very good

FUEL CONSUMPTION
Test conditions, mpg.......... 12.4
Normal cond., mpg............ 12-15
Cruising range, miles........ 200-250

DRAG FACTOR
Total drag @ 60 mph, lb........n.a.

CAMARO Z/28

continued

transmission does just this, keeping the engine up in the peak of its power band all the way through the gears. Perfect synchronization, low-effort shifting and reasonable gear noise are all characteristic of the Muncie unit. Unfortunately, the Z/28 suffers from the same malady as all Chevrolets, except Corvette, in its use of terrible shift linkage. The test car not only suffered from binding and lack of precision in shifting, but the linkage came loose during the test period and made selection of first and second gears nearly impossible. Tightening the shift rods brought the linkage back to its original state of barely acceptable operation. An enthusiast should install the excellent Hurst linkage assembly used on Pontiacs, or some equally good mechanism.

The Z/28 comes equipped with 15 x 6 in. rims, fitted with special E70-15 low profile tires. Wide Tread GT tires are Goodyear's latest, and feature steep cord angles for improved handling response. Cornering power is very high primarily due to the wide contact patch of these tires. Traction on takeoff is also good (the test car needed 3500 rpm at clutch engagement to keep the engine from bogging).

Our test car had power brakes, with discs at the front and drums in the rear. This is a mandatory option with the Z/28 package, and proved to be a good brake system. Maximum deceleration rate was 25 ft./sec.^{2}, and fade was negligible through CAR LIFE's standard eight-stop, 80-0 mph panic stop test cycle. This performance puts the Z/28 among the better domestic passenger cars, and matches the performance of the two Corvettes tested last month. For all-out racing, the Z/28 is available with discs all around, but this option requires a completely different rear axle housing. We tried a Z/28 with all-disc setup at the GM Proving Grounds at Mesa, Ariz., this spring and found them to offer fantastic fade resistance and even higher maximum deceleration than the standard disc/drum setup. However, four-wheel discs are sure to be an expensive and difficult-to-obtain option on the Z/28.

The test Z/28 had power steering. Chevrolet engineers have developed a system which affords low effort, fast steering and refuses to lose assist even during rapid maneuvering. Most domestic systems can be "beaten" during quick steering reversals. The Camaro always delivered consistent boost, except for some high effort which occurred only while the engine was idling.

ALTHOUGH TEST CAR did not have four-wheel discs, earlier drive at GM's Mesa, Ariz., test track with rear-wheel disc option (above) was totally satisfying.

Camaro handling, with standard Z/28 suspension, is impeccable on smooth surfaces. Slight understeer can easily be neutralized by application of small amounts of throttle. If oversteer is desired, to negotiate a tight turn even more rapidly, a sharp stomp on the accelerator pedal will force the rear end out to any degree the driver desires. Only Corvette, among domestic automobiles we have tested, affords this degree of agility and controllability. Steering response is well above other Ponycars, and cornering power is higher than any passenger car (except Corvette) that we've tested this year.

And the Z/28 is an easy car to drive very fast. Cornering does not require lightning-quick reactions, the Z/28 moving into each phase of its cornering performance with predictability and smoothness.

Only on rough surfaces did the Z/28 tend to lose its grip on the road. Rear axle hop on rough roads was noticeable—and objectionable. The front end washed out entering tight, bumpy turns, and directional stability over rough pavement was slightly skittery. On most highways and back roads, however, the Z/28 handled beautifully. ∎

EXTENDED TOURING in the Z/28 requires that the rear seat be used for luggage. Trunk cavity is small, and not shaped to accept standard luggage pieces.

Aerodynamic duo

Front: Camaro SS Sport Coupe. Rear: Corvette Sting Ray Coupe.

They're two of a kind. The fantastic, low-slung Corvette Sting Ray. And Camaro, The Hugger, the only car that comes even close. In styling, in handling, in performance. Both are aerodynamic from nose to deck, with Astro Ventilation, full door-glass styling, bucket seats, refined suspension and 327-cu.-in. standard V8s. You can order Vettes all the way up to 435 hp in a 427-cu.-in. Turbo-Jet V8. Camaros score almost as high: Cubes — 396, Horses — 325. Corvette's a tough act to follow. Buckle up a Camaro and see what we've done for an encore.

Camaro '68 CHEVROLET Corvette

A little goodie from the far reaches of the Chevrolet Sports Department that's just got to be an automatic contender.

Camaro Two-Step

by Eric Dahlquist

Practically nobody in Detroit even admits that there is a car called the Volkswagen but lo, just about the time VW announces their "Stick-Shift" automatic (see p. 70) Pete Estes, General Manager of Chevrolet and Vice President of GM, reveals the Torque-Drive, a 2-speed torque-converter transmission that "eliminates the clutch pedal, offers comparable economy to the regular 3-speed manual and has a list price of only $65." An automatic transmission for 65 bucks. That's not a bad deal, especially compared to the going beetle tab of $135 (that also includes a completely redesigned rear suspension). Well, when it finally got into the window sticker list it was really $68.65 but that's about as close as you can expect the Establishment to come.

In practice, the first thing you notice about the Torque-Drive-equipped Camaro is the steering — or lack of much of it — while negotiating the first turn and nearly knocking off an elderly lady and a row of parked cars. After a few miles and some more turns you'd be surprised how fast you adapt to the standard steering. But 5 turns lock-to-lock *is* a lot to ask — the '50 Chevy my brother Charlie had did better than that. This was supposed to be an economy Camaro, so it came with manual steering. We can't see why, though, in a relatively light car like this a faster ratio, along with some road feel, couldn't be supplied. The Chevrolet Zone Office neatly sidestepped the problem by installing a power assembly which we will have to admit is everything the standard box is not except cheap — add $84.30, please.

The Torque-Drive is the greatest thing since the cotton gin. It is easy to work, foolproof, fun and retains some of the physical involvement of manual shifting with almost none of the exertion. The closest thing it can be compared with is the old Powerglide because it is one — suddenly it's 1950. A cutaway drawing of the T-D shows it is indeed the familiar 2-speed automatic with the automatic valve body, governor, vacuum modulator, high-speed downshift mechanism and other trifles deleted. The whole deal is very clever, but it begs this simple question: Why, since it is very similar to a regular automatic, can this one be marketed for under $70 when the other is a $175 option?

Torque-Drive offers three gears: 1st (2.1:1), High (1.83:1) and Reverse. High is for normal driving, 1st, for extra dig and reverse, well, for reversing. Simple enough. The factory doesn't say you have to shift and you don't, but they've made it very easy if you want to. With almost any other column selected automatic you can name — and most of the console jobs as well — it is not difficult to overshift into neutral. This is a feat, if oft repeated at speed, that may lead to your undoing. With T-D, gear positions are very distinct and there is a little trick, positive-stop to keep you from up-shifting into neutral by accident.

Even with the big, torquey 250 CID 6 ($26.35 extra) the plan is to start from rest in low because the final drive ratio is a rather big 2.73:1. Not that the engine won't pull high, but it feels a little like squeezing into an overripe tomato and 1st is pretty lively at the stop lights. It comes as no great surprise that the Camaro 6 with or without automatic is not a prime candidate for the drag racers' sleeper-of-the-year award. Yet, an 18.70/74.19 mph at Orange County Raceway hinted that the installation of something around a 3.90 rear end ratio along with a quicker distributor advance curve and a set of exhaust headers would expand the immediate education of a few V-8 owners.

Walter Mitty racing potential or whatever, the real worth of the Camaro 6 automatic plainly is that it represents a great little integrated driving package — a welcome relief from the power-mad supercar whose acceleration response is so intense and overwhelming it takes on the character of a partially controlled ballistic missile. Not only that, but in a stripped machine like this you begin to see how superfluous such things as a console and deluxe buckets are when there is a noticeable increase of interior space without them. Sure, the brakes are not of the aircraft-carrier-arresting-hook-type and the car seems to float at around 95-100 mph but you can forgive these things because the machine is so well balanced otherwise. Not a wide-ovalled tire or a stiff suspension-that-feels-like-advanced-rigormortis-has-set-in anywhere and the little machine fairly flys through the esses. Even the fresh air ventilation worked well, allowing comfortable window-up draft-free driving and that is a first for an American car without air conditioning.

About the only negative item in construction was that your foot would sometimes catch on the underside of the wide brake pedal during quick stops. Economy as envisioned by the public relations gang was about 20 mpg but we knew better than this, expecting 14-16 and surprised at getting as high as 17.5. Price was just the opposite—we thought it would be modest but it wasn't. Somehow an economy car with a tab of almost three grand ($2901.90 f.o.b. Detroit) comes off badly in comparison with a comparably equipped VW for $800 less. This, plus the knowledge both vehicles will be of approximate value after a year or so of depreciation, is a hard pill to swallow. Just the big 6 and the Torque-Drive would come to $2683 which is more like it. Aw, but there's that darned steering that needs power. Come on, you guys in Flint how about a good, quick production manual? You want my brother Charlie to buy a bug or something? /MT

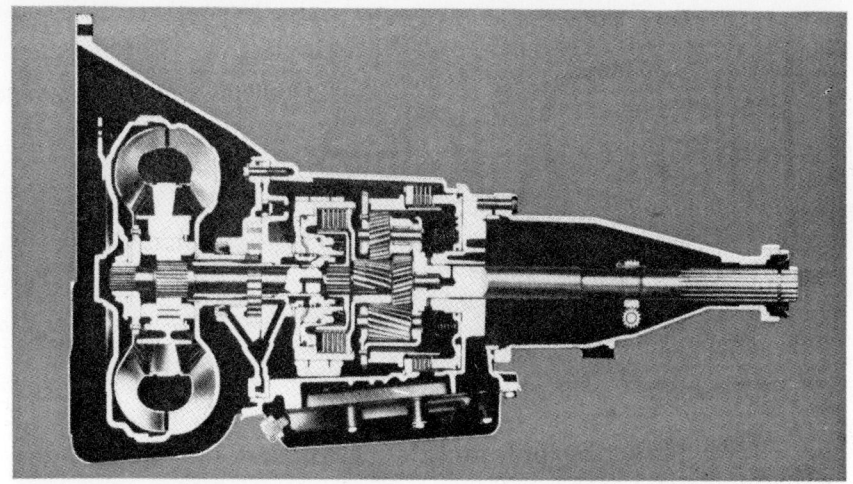

(Above) Lay a cutaway of the Torque-Drive over the Powerglide and it looks pretty much the same except that some of the parts are missing. (Below) Our car was Rallye Green — the soul of greenness — moss on an Irish castle. Unreal!

The 250 cube 6 has enough torque to bowl over banana trees and will even do a decent job on the highway scene.

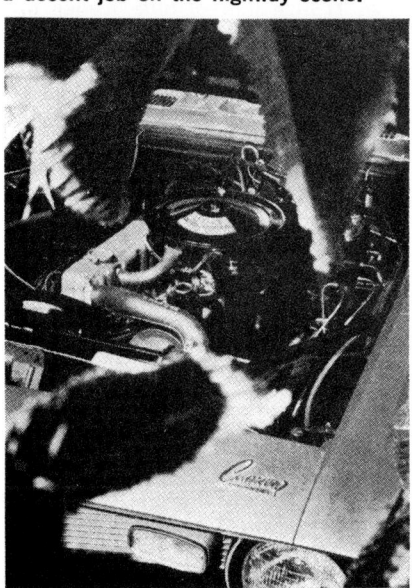

SPECIFICATIONS

Engine: Conventional OHV six. **Bore & Stroke:** 3.875 x 3.53 ins. **Displacement:** 250 cu. in. **HP:** 155 @ 4200 rpm. **Torque:** 235 @ 1600 rpm. **Compression Ratio:** 8.5:1. **Carburetion.** 1 Rochester single barrel. **Transmission:** 2-speed semi-automatic. **Final Drive Ratio:** 2.73:1. **Steering:** Semi-reversible recirculating ball nut. **Turning Diameter:** 41.0 ft. curb-to-curb. 4.8 turns lock-to-lock. **Tires:** 7.35 x 14. **Brakes:** 4 wheel hydraulic drum, dual system, 9.5-in. front and rear. **Suspension: Front:** Independent coil. **Rear:** Salis-axle with two single leaf springs. **Body/Frame:** Integral with ladder-type front section. **Dimensions, Weights, Capacities: Overall length:** 184.5 in. **Overall width:** 72.3. **Overall height:** 50.9. **Wheelbase:** 108.0 in. **Front track:** 59.6 in. **Rear track:** 59.5 in. **Road clearance:** 5.1 in. **Curb weight:** 3130 lbs. **Oil capacity:** 4 quarts. **Fuel capacity:** 18 gals. **Cooling system:** 12 quarts.

OPTIONS & PRICES

Retail price F.O.B. Detroit $2901.90. Options: 250 "6," $26.35; trim group, $42.15; AM radio, $61.10; white-wall tires, $31.35; Torque-Drive automatic, $68.65; power-steering, $84.30.

PERFORMANCE

Acceleration (2 aboard)

0-30 mph	5.5 secs.
0-45 mph	9.0 secs.
0-60 mph	13.6 secs.
0-75 mph	21.7 secs.

Speeds in Gears

1st	52 mph @ 4000 rpm
2nd	96 mph @ 4000 rpm
3rd	NA
4th	NA

MPH per 1000 RPM: 19.1 mph

Standing Start ¼-mile
74.19 mph, 18.70 secs.

Passing Speeds (1st, high gear)

40-60 mph	6.2 secs.
50-70 mph	8.5 secs.

Stopping Distances:

from 30 mph	32 ft.
from 60 mph	169 ft.

Mileage Range: 14 to 17.5 mpg

CAMARO Z/28

IT'S BEEN CALLED all sorts of things by different people: unmentionable names by Mustang drivers along the Trans-Am trail, a non-racing car by Chevrolet, a mini-Vette by admen and road testers.

Whatever you call it, the Camaro Z/28 can and does race and win. So for the person who wants the best-performing non-race car built in this country (and at a reasonable price, because nothing else retailing for $3400 base comes close), the Z/28 appears the only way to go.

Chevrolet rates the Z/28's 302-cid V-8 at a laughable 290 bhp. Four hundred comes much nearer the truth. This is a fistful of energy any way you dissect it. A 4-in. bore allows generously large valves, and the short 3-in. stroke keeps friction low and lets this

10 BEST TEST CARS OF 1968

engine wind to 7200 without missing a beat. By now every enthusiast knows that Chevy engineers took the 283 crankshaft and stuck it into the 327 block. Then by adding the full hop-up treatment—big 4-bbl. Holley on a smooth manifold, cavernous ports, huge valves, 346°-duration high-lift cam, cast-iron headers—the Chevrolet non-racing team wrought what amounts to a straightforward fairly simple and unsophisticated but highly potent, reliable powerplant.

We must stress here that the Camaro Z/28 isn't a car for everyone. We rate it most satisfying in its category, but that category is only for persons who put pure all-around performance ahead of pure transportation, tractability, a nice, smooth idle and lots of automatic conveniences. It takes a pretty determined driver to overlook the Z/28's little quirks—its lumpy idle, lousy gas mileage, poor low-speed torque, noisy engine, and stiff ride. But for the enthusiast those disadvantages are more than offset by

fierce acceleration, exuberant cornering, superb brakes, and the general feeling you get just sitting in a car that's gone all the way on everything.

In its high state of tune, torque for takeoff from rest comes in at about 3500 rpm. Set the tach needle on that figure and pop the clutch. Then, after the rear tires' tendency to glow subsides, the Z/28 takes hold, pulls 60 mph in 7.4 sec., storms through the quarter-mile in twice that (14.85) at 101.4 mph. Top speed with the standard 3.73 rear axle should be near 140 mph at 7000 rpm in high gear. That's assuming the engine will pull those tall revs in high gear and that everything stays together long enough to get there. We didn't try to squeeze this much from the car, but we managed to peg the needle on the admirably truthful 120-mph speedo in less than a half-mile distance.

Most U.S. cars that come with close-ratio 4-speed transmissions don't need them. Their rpm range is too short, so you stay in the intermediate

total engineering effort. All parts go with all others.

Among the most admirable components in this package are the brakes. Standard: power discs up front, drums in the rear. Optional: discs all around, but these should be considered only for all-out racing.

We've driven Z/28s both ways, and in either case the brakes again proved above the norm. Our test car's disc/drum setup also matched the two Vettes we tested this year (see page 28 for our estimate of the 427 Corvette), which means that this system gave us reliable, practically fade-free, straight-line deceleration through eight maximum-effort stops from 80 mph (1-min. intervals for cool down). That's a record very few domestic cars can match, especially those weighing 3700 lb. in road trim.

While quick, the 17:1 power steering ratio in our Z/28 isn't the fastest offered. There's a 15:1 ratio, too, but that would demand lots of attention at turnpike speeds. As it was, steering proved not only light and quick but also positive. With some power-steering systems, the driver can "beat" the power assist's response. There's a lag between moving the wheel and the time the front wheels turn. In systems like this, you have to anticipate fast corners. But this wasn't the case with the Z/28. When the steering wheel turned, the front wheels turned and so did the car.

The Z/28 Camaro didn't lean much. And going through flat, smooth corners on pavement can be pulled off in a normal understeering fashion or, by punching the throttle, in any degree of oversteer desired. You can hang the tail out most anywhere because there's so much power on tap. Okay—this is fine on hard, smooth surfaces. But in the rough, where there are bumps, ridges, gravel, loose sand, or anything else, the suspension's stiffness creates a good deal of sideways hopping— much too much for comfort. All domestic cars with firm suspensions do this, so we can't fault the Z/28 too much except to say that perhaps something like the Corvette's independent rear suspension might help.

Still and all, the Z/28 offers a number of blessings the Vette doesn't. First, there's the indisputable price advantage. Then there's room for four passengers and a trunk that holds at least the bare necessities of cross-country travel. There's the knowledge and the feeling of maximum performance from a small-cid engine. This car draws nearly everything that can be drawn from a moderately sized, lithe, muscular package. It still tucks you back into the bucket whenever you punch it above 3000 rpm. And more than that, the car's other systems— brakes, steering, suspension—match its power. The Z/28 comes off as a highly satisfying vehicle, able to hold its own against any competition.

gears less than a second in many cases. But in the free-winding Z/28, its standard Muncie close-ratio box does an admirable job.

This car *needs* this transmission. With adequately complementary shift linkage (which our test car didn't have), the Muncie remains a delight to use, beautifully suited to any track or road situation.

The Camaro Z/28 is now being sold as a street machine. Last year (1967) this package went only to drivers who'd race it. Considering this background, the Z/28 does all right on the street. Driving down the highway, there's a fair amount of jostle inside the car. This comes from a firm suspension system—stiff springs, stiff shocks, stiff anti-roll bar. But that's what you'd expect, and it makes for extremely good, fast cornering power. The Z/28's ability to stick through turns remains among its finest virtues; far superior among Chevrolets to anything but the Corvette. Helping in this department are Goodyear's special E70-15 low-profile Wide Tread GT tires. These put a monstrous patch on the road.

Often Detroit stuffs a high-bhp engine into a relatively small car and lets it go at that. Not so with the Z/28. Happily, this car represents a

SPECIFICATIONS

Wheelbase, in.	108.0
Overall length, in.	184.6
width	72.3
height	50.9
No. of passengers	5
Frame type	Unitized
Front suspension: Independent by s.l.a., coil springs telescopic shock absorbers.	
Rear suspension: Hotchkiss live axle, multileaf springs, staggered shock absorbers.	
Steering: Integral assist recirculating ball gear, parallelogram linkage behind front wheels.	
overall ratio	17.0:1
turns, lock to lock	2.8
turning circle, ft. curb-curb	37.0
Curb weight	3355
Brakes: Disc front, cast iron drum rear.	
dia. x width, in. F/R	11.0 x 2.21/9.50 x 2.0
total swept area, sq. in.	332.4

Tires: (Goodyear Wide Tread GT) size	E70-15
Engine: ohv 90° V-8	
Bore x stroke, in.	4.00 x 3.00
Displacement, cu. in.	302
Compression ratio	11.0:1
Rated bhp @ rpm.	290 @ 5300
Clutch: Single dry disc, diaphragm-type pressure plate.	
diameter, in.	11.0
Transmission: Four-speed fully synchronized.	
Gear ratio 4th (1.00:1)	4.10:1
3rd (1.27:1)	5.21:1
2nd (1.64:1)	6.72:1
1st (2.20:1)	9.02:1
Lb./bhp (test weight).	12.7
Mph/1000 rpm (high gear).	18.4
Engine revs/mile (60 mph)	3260
Piston travel, ft./mile.	1630
CAR LIFE Wear Index.	53.2

ROAD TEST RESULTS

Speedometer reading at 30 mph	30.6
Speedometer reading at 60 mph	59.3
Top speed (7200)	133
Acceleration 0-30 mph, sec.	3.5
0-40 mph	4.5
0-50 mph	5.7
0-60 mph	7.4
0-70 mph	8.9
0-80 mph	10.5
Standing ¼ mile, sec.	14.85
Speed at end, mph	101.4

Passing, 30-70 mph, sec.	5.4
Braking: Maximum deceleration rate ft./sec²	25
No. of stops from 80 mph (at 60-sec. intervals) before 20% loss in deceleration rate	8 no loss
Control loss: Slight. Overall brake performance	very good
Fuel consumption under test conditions, mpg.	12.4
Normal cond., mpg.	12-15

Z is for "Zap!"

Translation: a 302 V8 with mechanical lifters, hi-performance cam, aluminum intake manifold, Holley 4-barrel.

Plus: multi-leaf rear springs, heavy-duty shocks, new white-lettered tires on 15 x 7 wheels.

And a Hurst shifter for the 4-speed.

While you're at it, why not add the new 'Vette type 4-wheel disc brakes?

By now you know the mean streak isn't just painted on—it's built in.

CHEVROLET

Putting you first, keeps us first.

We've got a mean streak.

GM

MARK OF EXCELLENCE

Z/28 Camaro.

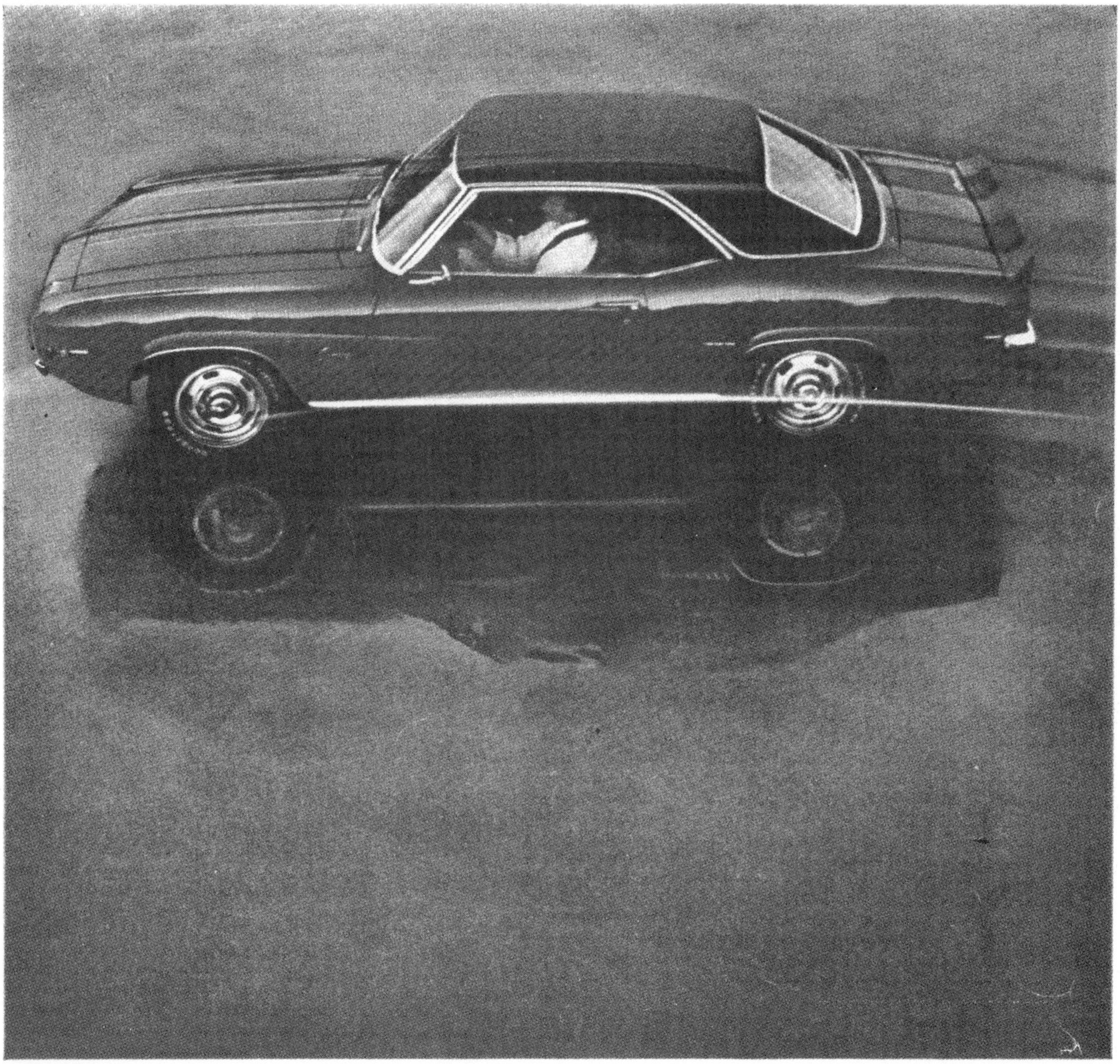

Up until last year there was but one American built automobile available with four wheel disc brakes. Now there are two and they're both sold under the same banner — Chevrolet. A late comer into the Ponycar picture, Camaro, according to VP Pete Estes, should overtake Mustang's lead sometime in 1969.

Although really not in racing, Chevrolet offers about the hottest complete package for quick over the road touring in their Z-28. Except in the straights it will give sister Corvette a helluva run for her money and carry four passengers to boot. For a company not in racing a Z-28 is about as close to an over the counter racer as you can get. Then there's this guy named Penske and another named Donohue who worked one over and showed the rest of the Trans-Am boys the short way around the nation's race-tracks.

Shortest of the Ponycars at 186 inches, the Camaro offers buyers two sixes and three V-8 sizes for the biggest engine spread available. Match either with a three or four-speed manual box or a two or three-speed automatic and a wide range of axle ratios and you can brew your own brand of performance or economy. Horses range from 140 for the 230 CID six to 325 for the 396 V-8 and bodies come with hard or soft top. New items include 7-inch rims (8 on SS models), variable-ratio power steering, headers and hood scoops that breathe.

With the six you can even get a new, less expensive automatic two-speeder for about $80, and, meanwhile back at the proving grounds, they're experimenting with a neat little rig with two four-barrels on a special intake manifold that launches you into the high three digit figures before you can say, "We're not really in racing."

Probably the 350 CID V-8 will be the hot number for 1969 and owners can dress it up with bright orange colors, spoilers front and rear, and hidden headlights.

Camaros won't drive you 'absolutely mod' this year, but they are cars meant to be driven. With the right choice of suspension, engine, transmission, axle ratio, and those beautiful four wheel disc brakes, the real sports car owners will have to sit up and take notice, usually from behind.

From the 350 CID Camaro you can expect quarter miles in the 16-17-second range between 83-90 mph and even the six will top the century mark on a good day.

CAMARO

Camaro
Data in Brief

DIMENSIONS

Overall length (in.)	186.0
Width (in.)	74.0
Height (in.)	51.6
Wheelbase (in.)	108.0
Track front (in.)	59.6
Track rear (in.)	59.5
Turning diameter (ft.)	37.0
Fuel tank capacity (gal.)	18.0

WEIGHT, TIRES, BRAKES

Weight (lbs.)	3260
Tires	F70 x 14
Brakes, front & rear	drum

ENGINE

Type	V-8
Displacement (cu. in.)	350
Horsepower	255

SUSPENSION

Front	independent coil
Rear	single leaf

HUGGER-STRIP TEASE

The Camaro SS 396 promised fierce performance, but its unrefined rear suspension wouldn't let it deliver.

ADHESION LIMIT of the Camaro has just been reached. Weight transfer is lifting the inside wheel off the ground and putting more weight on the outside rear tire than the tire can handle. The inside front wheel isn't doing much, either.

PHOTOS BY SCOTT MALCOLM

CRANKED INTO a corner at the limit, the Camaro 396 develops severe understeer. The front wheels are turned against the weight of the car, but it isn't doing any good; body roll has tipped them the wrong way. The big engine unbalances the car.

MAKE A LIST. Help Chevrolet strike back, with a big-engine Camaro. The 396-cid V-8, in 375-hp tune. The SuperSport package —power-assist brakes with front discs, stiffer springs and firmer shock absorbers, wide wheels and big tires. A four-speed stick with Hurst shifter. Quick power steering. Cold-air hood, with inlet in a high-pressure area.

Add it up, and it doesn't add up. The combination totals less than the parts. When Chevrolet made that list, something was left out. The rear suspension is plain vanilla, weak linkage between axle and car, and it drags the 396 Camaro down to the level of just another Camaro.

The rear axle is a live axle, meaning just what it sounds like. At the mere suggestion of work, the axle leaps and hops, judders and bucks, like the kid who bawls before the switch gets close. Starting, stopping or turning, whatever the rest of the car wants to do, the rear suspension won't let it do it.

Look at the acceleration times. The 396 puts out at least its rated power. Performance gearing, a shifter that moves as fast as the fastest hand, and the quarter-mile E.T. is 14.77 sec. Why? Because axle judder increases in direct proportion to traction. Light the tires, and the speedometer needle spins while the car just sits there. Do it right —catch the best place on the power curve, where there's just enough poke to belt off the line with a trace of spin —and the axle bangs on its mounts like Skinflint John come for the rent.

With each fast shift there's the thump of Moby Dick versus whaleboat. We wouldn't bash our own cars like that, so we settled for kindness and times that are in the Supercar class, but don't show what the engine would do if it could.

The brake tests went the same way. The system is a good one, better than the deceleration figures show, and the figures are pretty good anyway. Effort was easy to control. The testers could muster enough pressure to make all four tires bite; and crash. The Man wants to get our attention again. The deceleration figure for the first stop actually was the second try. On the first, the thrashing and leaping tore the decelerometer off the windshield, like King Kong dealing with those pesky

65

INSTRUMENTATION was complete. The tachometer can be seen instantly, and the gauges on the console are available for occasional consultation.

CAMARO SS 396

continued

airplanes. With the drums heated, the rear brakes faded; but still, they skipped or locked while the front tires were trying to get the car stopped. Pushing past the soft side of maximum traction locked both rear wheels.

Cured the judder, but the rear made signs of wanting to lead.

That's the polite part. For the handling discussion, we get rough. Big engine = pile of weight up front. Understeer again. Cruise into a fast turn at speed, and the front outside fender dips while the tires slant over and the mass of engine shoves the front wheels toward the outside of the curve. And the power comes on. When it does, the rear tires fling their feet in the air. In-

stant sideways. The driver dials in correction on the very light steering. The best he can do is back off and catch the car before it spins—to the inside—or slides off the track—to the outside. We never did manage to power the Camaro out of a turn, the way we could with either the 427 Corvette or the 327 Chevy II tested last year. When the tires lose their side grip, the forward grip goes with it. Power doesn't move the Camaro forward, it rotates the rest of the car around the engine. If an instructor told his engineering class to design a car with massive initial understeer, severe final oversteer and a general feeling of wishy-washy before and during the transition, they'd stalk from the classroom in protest. But Chevrolet did it, presumably without trying.

The Trans-Am Firebird we drove last month had the same basic body/chassis, similar suspension and an equally big engine. It had so much lateral grip in back that we couldn't hang the tail out; more power shoved the car off course, all right, but in a dependable straight line. Just goes to show, all domestic cars aren't alike. When they are, maybe there'll be a balance between the two extremes displayed by the Camaro and Firebird.

Off the test track, the 396 Camaro

1969 CAMARO SS 396

CHASSIS/SUSPENSION
Frame type: Unitized.
Front suspension type: Independent by s.l.a., coil springs, telescopic shock absorbers.
  ride rate at wheel, lb./in......n.a.
  antiroll bar dia., in..........0.688
Rear suspension type: Live axle, multileaf springs, staggered shock absorbers.
  ride rate at wheel, lb./in......n.a.
Steering system: Integral assist recirculating ball gear, parallelogram linkage behind front wheels.
  overall ratio..............17.0:1
  turns, lock to lock..........2.8
  turning circle, ft. curb-curb.....37
Curb weight, lb................3490
Test weight...................3790
Distribution (driver),
  % f/r................. 59.3/40.7

BRAKES
Type: Disc front, drum rear.
Front rotor, dia. x width, in.11.0 x 2.5
Rear drum, dia. x width.....9.5 x 2.0
  total swept area, sq. in......332.4
Power assist
  line psi at 100 lb. pedal........790

WHEELS/TIRES
Wheel rim size................14x7
  optional size................14x6
  bolt no./circle dia. in......5/4.75
Tires: Goodyear Wide Tread.
  size................F70-14
  normal inflation, psi f/r.....24/28

ENGINE
Type, no. of cyl.................V-8
Bore x stroke, in.......4.094 x 3.76
Displacement, cu. in.............396
Compression ratio..........11.0:1
Fuel required.............premium
Rated bhp @ rpm.......375 @ 5600
  equivalent mph.................111
Rated torque @ rpm.....415 @ 3600
  equivalent mph.................72
Carburetion: Holley 1x4.
  throttle dia., pri./sec.....1.56/1.56
Valve train: Mechanical lifters, pushrods and overhead rocker arms.
  cam timing
  deg., int./exh.......44-92/86-36
  duration, int./exh........316/302
Exhaust system: Dual, reverse flow muffler.
  pipe dia., exh./tail.....2.50/2.25
Normal oil press.@ rpm. 50-75 @ 2000
Electrical supply, V./amp......12/37
Battery, plates/amp. hr........66/61

DIMENSIONS
Wheelbase, in.................108
Track, f/r, in.................60/60
Overall length, in..............186
  width.....................74
  height....................52
Front seat hip room, in........21 x 2
  shoulder room.................57
  head room....................37
  pedal-setback, max.............41
Rear seat hip room, in...........55
  shoulder room.................54
  leg room....................29
  head room....................37
Door opening width, in.........40
Trunk liftover height, in........32

PRICES
List, FOB factory............$2743
Equipped as tested...........$4294
Options included: 396-cid/375-bhp, $316; SS package, including power disc brakes, stiff suspension, 14x7 wheels, F70-14 tires, $296; power steering, $95; tach and console, $95; custom interior, $110; ZL-2 hood, $79; radio, $61; 4-speed trans., $195.

CAPACITIES
No. of passengers................5
Luggage space, cu. ft...........8.3
Fuel tank, gal................18.5
Crankcase, qt..................4
Transmission/dif., pt........3/3.5
Radiator coolant, qt.............23

DRIVE TRAIN
Clutch type: Single dry disc, diaphragm-type pressure plate
  dia., in....................11.0
Transmission type: Four-speed fully synchronized.
Gear ratio 4th (1.00:1) overall.3.73:1
  3rd (1.27:1)......4.74:1
  2nd (1.64:1)......6.02:1
  1st (2.20:1)......8.21:1
Shift lever location: Console.
Differential type: Hypoid, limited slip.
  axle ratio..................3.73:1

still was quirky. On a crowned road, or one with dips and ridges, the rear end walks around. The driver constantly corrects, this way and that. It's a subjective, rather than scientific, criticism, but the light steering feels imprecise, and never tells the driver enough. On wet pavement, the brakes need care. Again, the light rear end dances around. One tester got slightly sideways at two consecutive stop lights. Fuel mileage was on a par with the Supercars of equal gearing (and better acceleration times), while the gears and big engine couldn't be used. Better to pick the big and downhill gears or score at grudge night with a smaller engine that won't overtorque the rear axle.

Sorry about all this. The 1969 Camaro has several nice touches that the critic shouldn't overlook. The front bumper is plastic, a rub-off from Pontiac, and an attractive one. The cold-air hood, code-named ZL-2 by the same guy who comes up with Z/28, L-88, ZL-1, LT-1 and so forth, doesn't catch the eye the way the forward-facing scoops do, but putting the inlet just ahead of the windshield base means it pulls in more air. The emissions snag is whipped by a valve that shuts off the cold air until the throttle is 80% open.

INTAKE FOR the cold-air system is at the base of the windshield, where it benefits from high pressure at speed. New option is code-named ZL-2.

Inside, the seats are comfortable, and the control placed within reach. The instrument option puts the tachometer high in the panel, to the right of center. The gauges go in the console. They only need an occasional glance. We'd prefer all the instruments in the panel, but if they won't fit, the Camaro compromise is one way out.

We laughed when one staffer discovered that the fenders on his import don't match, that the left front isn't the same as right front. Bert did one side, 'arold the other, and Bert and 'arold don't speak. Couldn't happen here. We don't have craftsmen, we have machines, computers, engineers, laboratories and proving grounds.

So Chevrolet builds a car with a rear that doesn't match the front. We don't know whether to suggest that Chevrolet come out with a set of control arms, or just let Bert and 'arold laugh last. ∎

CAR LIFE ROAD TEST

ACCELERATION & COASTING

MPH

ELAPSED TIME IN SECONDS

CALCULATED DATA

Lb./bhp (test weight)	10.1
Cu. ft./ton mile	182.1
Mph/1000 rpm (high gear)	19.9
Engine revs/mile (60 mph)	3010
Piston travel, ft./mile	1885
CAR LIFE wear index	56.7
Frontal area, sq. ft.	21.4

SPEEDOMETER ERROR

Indicated	Actual
30 mph	30.0
40 mph	40.8
50 mph	50.8
60 mph	61.4
70 mph	71.8
80 mph	81.6
90 mph	92.0

MAINTENANCE

Engine oil, miles/days.....6000/120
oil filter, miles/days.....6000/120
Chassis lubrication, miles.......6000
Antismog servicing, type/miles..replace PCV valve/12,000, tighten belts, tune check/12,000
Air cleaner, miles.....replace/24,000
Spark plugs: AC 43N.
gap, (in.)..................0.037
Basic timing, deg./rpm..4 BTDC/900
max. cent. adv., deg./rpm..32/5000
max. vac. adv., deg./in. Hg..12/12
Ignition point gap, in...........0.019
cam dwell angle, deg........28-32
arm tension, oz............19-23
Tappet clearance,
int./exh............0.024/0.028
Fuel pressure at idle, psi.......5-6.5
Radiator cap relief press., psi.....15

PERFORMANCE

Top speed (6300), mph	126
Test shift points (rpm) @ mph....	
3rd to 4th (6300)	99
2nd to 3rd (6300)	76
1st to 2nd (6300)	58

ACCELERATION

0-30 mph, sec.	2.6
0-40 mph	4.0
0-50 mph	5.1
0-60 mph	6.8
0-70 mph	8.3
0-80 mph	10.4
0-90 mph	12.5
0-100 mph	15.6
Standing ¼-mile, sec.	14.77
speed at end, mph	98.72
Passing, 30-70 mph, sec.	5.7

BRAKING

Max. deceleration rate from 80 mph
ft./sec./sec....................27
No. of stops from 80 mph (60-sec. intervals) before 20% loss in deceleration rate................7
Control loss? Extreme (rear axle hop). Overall brake performance
....................fair (see text)

FUEL CONSUMPTION

Test conditions, mpg.	8.5
Normal cond., mpg.	8-12
Cruising range, miles	148-222

Make your Camaro Handle

Modify your Camaro's suspension system with heavy duty parts from the Factory.

So you're the proud owner of a Camaro? And you've decided to go racing with it? Great. How about letting us help you get it ready? Oh, you already have a high performance engine all blue-printed and full of go. Good for you. What about the chassis? You *weren't?* You'd better; follow along as we work our way through the various chassis components. The specifications and suggestions are intended to assist in preparation of a Camaro for road racing as well as drag racing applications. As you probably know, regulations vary somewhat regarding what *can* and what *can't* be done as far as a particular sanctioning organization is concerned; therefore, you'll need to check the rules books to see which of the procedures described are allowed for the particular application you have in mind.

Let's dismantle the front subframe first and get to work. Remove it from the car and thoroughly clean (preferably by sand-blasting) the subframe. All seams should be checked and completely welded if any weak spots are found. Reinforcement of the spring seat and upper control arm mounting bracket is recommended (remember to check those regulations we mentioned earlier). The shock absorber clearance hole in the upper spring seat should be cut out to insure adequate clearance around the shocks.

The rubber body mounts should be removed. Aluminum or steel spacers should be used to space the subframe in its normal position (relative to the unit body). The metal spacers eliminate flexing between the subframe and body, thereby improving handling and safety. For road race applications, bolting or welding the subframe solidly against the body is not allowed.

Next come the body and roll cage preparation. Remove as many unnecessary components as the sanctioning organization will permit; the objective is to achieve minimum weight, you know. Reweld or braze all body areas subject to stress. The roll cage will be attached above all rear spring mounting points so that reinforcements required to mount the cage will also reinforce the body at points of highest load. Fender wells can be reworked to accommodate the wide profile tires, and you should keep in mind that objective to decrease total car weight as much as possible. Don't get carried away on the fender-well modification to the extent that the rear wells are adversely affected. They provide a significant portion of rear body rigidity and should be kept intact. Widening in this area should maintain fender-well continuity.

A *complete* roll cage (not a bolt-in type) should be constructed and welded into the car. It should have four upright tubes (one each ahead of and behind the driver on both sides of the car). The rear uprights should tie into the body over the forward rear spring mounts. The front tubes should pass through the floor and be welded to the subframe. Tops of the uprights should be joined across the car at the top of the windshield and behind the driver's head in a continuous hoop with smooth bends. Horizontal connecting bars should join the two hoops at each side of the top.

Two rear braces should be installed from the rear of the cage to tie into the body over the rear spring shackle mounts. Two forward braces should tie into the subframe just behind the front upper control arms. To further stiffen the cage and prevent flexing under high chassis loading, triangulating braces are a worthwhile addition. A protective brace can also be installed along side the driver (between the door and seat).

The portion of the roll cage immediately behind the driver may have to be a specified diameter and wall thickness, depending on that sanctioning organization we keep mentioning. If you're tempted to forego the procedures described above and try the bolt-in approach, keep in mind that the object of having a roll cage is to increase safety and body-to-subframe rigidity. These will both improve with a sturdy cage structure.

Front suspension mounting points cannot be changed, according to the rules of most organizations. All rubber control arm bushings should, however, be replaced with bronze, aluminum, or high density plastic bushings to eliminate suspension compliance under cornering loads (road racing

Multitude of '69 pieces are directly applicable to earlier Camaro units. Most of the parts are specially suited to drag-strip application. . .especially front and rear spring rate combinations.

applications) and provide better tire contact from both front wheels. Keep in mind that these bushings are subject to wear and should be designed to keep clearance from developing as wear progresses.

High quality ball joints and tie rod ends are recommended for serious competition, as well as Tufftride heat-treated knuckles. A retaining strap across the bottom of the lower ball joint/lower control arm assembly or tack welding is recommended to retain the assembly (in addition to the press fit).

Anti-sway bars are available as optional dealer equipment in sizes up to 1 1/16-inch OD, and these bars in production rubber mounting bushings, with production rubber cushioned links, have proven satisfactory in competition. Large diameter bars can bind up in production chassis mount rubber, so if you're going for the bigger sizes, give some consideration to this area. The front ride height should be about 9-1/2 inches from the ground to the inner forward A-arm bolt centerline. Make sure enough "bump" travel is available at this point.

Complete package of pieces even include Trans-Am orientation, as you can note here in the highly-controversial No. 13 Smokey Yunick Camaro of 1967

Although not a direct factory suspension option, relocation of the battery to a point in the trunk can influence overall handling and/or weight transfer characteristics.

Rear spring options offer a good variety of rates. The forward spring eye bushing should be made of bronze or aluminum rather than rubber to help position the rear axle properly and reduce axle tramp during braking.

The rear axle should be attached to the springs with U-bolts of the heavy duty type. Rear body height can be adjusted by use of lowering blocks or via reaching of the springs at a spring shop. Rerolling of front spring eyes below (rather than above) the main spring leaf will improve rear axle tramp and permit removal of lowering blocks. Whatever the method, at least three inches of bump travel (measured from the axle top to the bottom of the subframe) should be allowed for severe jounce.

Rear roll-rate adjustments fall in the "pay your bucks and pick your favorite" category. Some chassis builder/drivers prefer anti-roll bars; others use springs. Pan-hard rods, Watts linkages, and traction bars or radius rods are all employed for this function. Some successful competitors use none of these extra rear axle components. The staggered shock absorber mounting, rerolled spring eyes and hard forward spring bushings do a fairly effective job of controlling power and brake hop or axle tramp.

For general race applications, 500 lb/in spring rate with 7/8-15/16 inch or 1 inch sway bar up front and 250 lb/in rear are a good starting point. You high-bank fans will need to up the spring rates to 725 in front and 300 in back.

As a rule of thumb, any time the front end pushes or understeers excessively, the front roll rate should be decreased by using softer springs or a smaller sway bar. Another alternative is to increase the rear roll rate by installing stiffer rear springs (or sway bar, if one is used in the back). If the car oversteers or tends to spin out too readily, softer rear springs or higher front spring rates and/or sway bar is needed. Understeer control to a degree can be reduced by changes to the camber (more negative camber to reduce the understeer). The most satisfactory balance for handling and maximum speed on road courses is a slightly understeering condition which becomes oversteer after power is applied. It should also be kept in mind that spring rates should prevent bottoming out of suspension and shock absorbers during heavy chassis loadings (such as corners); severe, immediate oversteer or understeer and possible loss of control can result. Suspension travel can be improved by trimming or removing bump stops.

Heavy duty racing shocks will be needed to provide satisfactory, safe handling, such as those manufactured by Monroe and Koni. A number of good shocks are marketed which offer various stiffnesses and jounce/rebound percentages, some utilizing integral progressive bump stops to limit suspension travel on jounce.

If you plan to use as many optional service pieces as possible rather than mixing and matching, the production disc front and drum rear brakes on Z-28's can be modified for limited competition where repeated, hard braking is not necessary. Heavy duty front brake calipers (#5468886 and #5468887) allow use of flanged, molded Corvette front brake pads (#5468882) on the 11-inch diameter, one-inch thick Camaro front rotors. Sintered iron rear brake shoes (#3830635) should be installed on the drum brakes, and the complete automatic brake adjusting mechanism should be switched side to side. The adjusting mechanism should be mounted on the forward shoe, which will cause the adjuster to operate when brakes are applied under forward motion. The star wheel mechanisms will also have to be switched. The backing plate and any other allowable brake ventilation achieved by drilling in order to cool the brake components is recommended.

A Chevelle disc-drum master cylinder (#5468811) will increase master cylinder capacity

Heavy-duty sway bars can be further "tuned" to a given car by the addition of adjustable A-arm-to-sway-bar links. Effective pre-load can then be set to any desired level within the range of the sway bar being used. Lift-pads (at the upper A-arms) of the type shown are not a factory option but can be bought at specialty equipment stores. Installation provides slightly better-than-stock wheel well clearance and improved weight transfer. Nose-up attitude of unit-body Camaro is quite critical, so try to maintain at least a 1-3 degree positive rake static attitude (front to back).

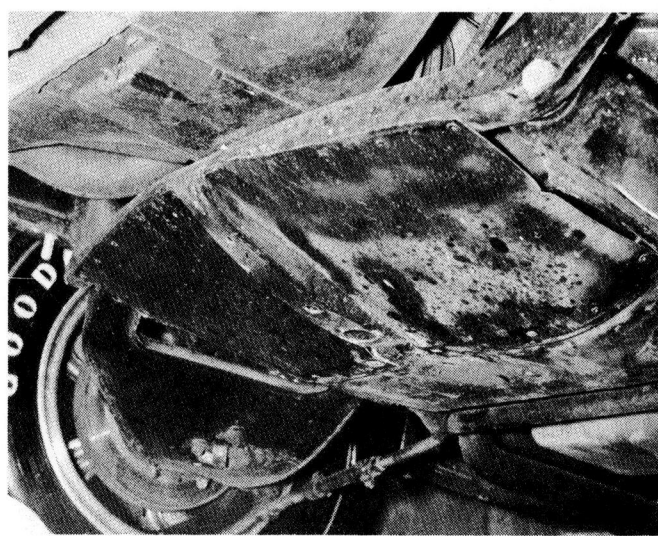

Addition of sheet-metal to underside of front cross-membering will aid under-car air-flow and assist in keeping the nose of the car down during high-speed operation.

Home-made spring clips help prevent axle housing wrap-up (torque reaction) during hard acceleration. Placement of these clips is best determined by trail and error. It's possible to make the spring too restricted,

Safety tabs (not a factory option) will temporarily hold a wheel/axle combination in tack following axle breakage. Without them, you could lose the car from a sudden and unexpected turn. Much cheaper than safety hubs.

for the front brakes, and shot-peened axle shafts (#3927508) are recommended for improved durability.

For all-out racing applications, complete front and rear discs which utilize heavy duty Corvette rotors, calipers, and pads are the way to go (or stop). Brake kits (#3947050, RH and #3947049, LH) utilize production Camaro knuckles and steering arms.

Rear disc brakes are available as part of a complete axle assembly (#3945131) with a 3.73:1 ratio and Positraction. With these brake components, master cylinder #5463751 should be used. It's the same as the aforementioned Chevelle equipment but has the residual pressure check valve removed from the rear pressure outlet.

Now that we've run through the parts list, let's consider some installation and maintenance tips. The caliper or rotor should be shimmed to center the rotor in the caliper. Front wheel bearings should be adjusted fairly tight. There should be no rotor wobble or lateral runout, which could cause brake pad "knock-back."

Rear rotor runout should be checked and eliminated and rear axle end float should also be reduced as much as possible. Control of end float on the rear axle is governed by the thickness of the

axle retaining "C" locks and the thickness of the Positraction clutch shims between the pinion side gears and differential case. End float should be adjusted to 0-.015 inch. After run-in it may be necessary to reshim these areas to maintain the tolerance.

Frequent brake line bleeding is recommended to keep all air out of the system. A 15-30 psi power pressure bleeder is helpful for this operation. Tapping the calipers with a plastic or rubber mallet will help loosen air bubbles. Brake lines should be routed so as to prevent loops or high points that can trap air and prevent complete bleeding.

The shackle between the brake rod and the brake pedal should be adjusted so that the master cylinder bottoms before the brake pedal contacts the floor. It may be necessary to trim or raise the upper brake pedal stop to permit the master cylinder to return to its closing stop. Open both master cylinder bleeder screes and depress the brake pedal to check out master cylinder travel. Rear brake lockup or rear axle tramp under hard braking conditions can be prevented by installation of an adjustable proportioning valve in the rear brake line. This valve is used on heavy duty Corvettes (bracket #3910796, proportioning valve #3878944, and brake pipe #3904972) to limit maximum rear brake line pressure.

Rear-view of tabs shows method of installation. Material can be flat-strap of 1/4 x 1-inch dimension. Attachment is with short hex-headed capscrews.

Safety-loop is a "must" in unit-body cars such as the Camaro. Without one, front U-joint loss generally means running over the driveshaft with a move somewhat akin to a poll-vaulter. . .and the results are far more spectacular.

A slight amount of rear tire clearance can be gained by addition of spring shackle extensions. The point to remember is that you can elevate the car to a point where additional height doesn't improve tire clearance but causes a center of gravity shift addendant with a "shaky" car. Raise the car only to a point that's needed, and stop.

Only the best quality brake fluid should be used for high temperature resistance. Fluid should not be left uncovered or reused. Delco Supreme 550 G (#5464591 in gallon cans) is available through United Motors Service outlets and has been found satisfactory.

A foam-filled bladder tank (installed inside the trunk) is definitely recommended to take care of fuel storage. Sanctioning organizations are requiring this equipment for safety reasons. Tanks of this type are manufactured by Goodyear, Firestone, and Don Allen Company.

The largest production radiator available for the Camaro has a 2.7-inch core (#3016688). Four- and five-blade solid as well as viscous drive fans are available, also. Oil cooling radiator #3157804 is recommended for competition applications. An oil cooler adapter is required to bolt to the cylinder block in place of the production oil filter. A remote oil filter is also necessary, installed into the line returning cooled oil to the engine. These oil cooler and filter adapters are available from several high performance parts manufacturers such as TRACO Engineering. Oil cooler lines should be a minimum of 1/2-inch ID, and the cooler will fit beside or in front of and beside the water radiator.

The production Z-28 oil pan and engine tray baffle have been satisfactory with the 302 cubic inch performance engine in racing applications. This pan has been run as much as two quarts overfull with no harmful effects. Overfilling by one quart is recommended.

There are, as you probably know, wheel-width limitations imposed by the various organizations. Magnesium wheels (preferred) or aluminum counterparts are available in the acceptable widths by several manufacturers in the Chevrolet five-bolt pattern.

To retain maximum rear tread width with minimum fender rework, .20-.25 inch positive offset is recommended. For the front wheels, a negative offset of the above amount will improve wheel bearing life, provide easier steering, and require a minimum of fender rework.

Improved durability will be obtained by use of 1/2 x 20 x 1 3/4 inch wheel stud bolts (#3849110). The splines will have to be shortened to .40 inch for rear hubs. Keep in mind that rear brake disc should not ride on the splined section of the wheel stud bolts. Special lug nuts to be used with three-inch long 1/2 x 20 lug bolts are available from American Racing Equipment in Brisbane, Calif. GM wheel stud bolts (#3819780) are 2 7/8 x 1/2 x 20 and may need to be shortened in the spline to fit the hubs you are using. These bolts and lug nuts facilitate quick wheel changes with power wrenches.

Racing tires are an article in themselves, and we'll let you decide which sticker you want on your windshield or whatever.

Let's balance out this consideration of nuts and bolts and chassis components with a brief discussion of front end geometry. Front wheel camber should be 2-3½ degrees negative, caster should range from 3-5 degree positive, and toe-out should be 1/16-1/8.

The chassis should provide approximately equal weight on both rear wheels with the driver seated in his position. Various length rear spring shackles can be used to accomplish this weight distribution, along with shimming or trimming the front springs as required. Anti-roll bars should be unhooked during this measurement and adjustment procedure, and the chassis should not be preloaded during the reconnection of the anti-roll bars.

With these modifications incorporated, the Hugger is ready for the road. Wanta drag to the first turn?

Z-28 CAMARO HEAVY DUTY CHASSIS PARTS

PART NO	DESCRIPTION	SPECIFICATIONS
STEERING		
3916236	Knuckle—steering	Tuftrided
9748406	Stud and seal—upper cont. arm ball	Tuftrided
3875067	Stud and seal—lower cont. arm ball	Prod. shot peened
3930030	Socket assembly—tie rod, outer	Shot peened
3930028	Socket assembly—tie rod, inner	Shot peened
3958493	Rod—steering relay	Shot peened
3916237	Arm—steering knuckle, L.H.	Shot peened
3916238	Arm—steering knuckle, R.H.	Shot peened
7806396	Gear Assembly—steering (RPON 44)	20:1 ratio
SPRINGS		
3948989	Spring assembly—front	507 lb/in rate
3948406	Spring assembly—front	561 lb/in rate
3935784	Spring assembly—front	615 lb/in rate
3935785	Spring assembly—front	723 lb/in rate
3948988	Spring assembly—front	777 lb/in rate
3947503	Spring assembly—front	864 lb/in rate
3948986	Spring assembly—rear	200 lb/in rate
3948985	Spring assembly—rear	250 lb/in rate
3953673	Spring assembly—rear	300 lb/in rate
3935786	Spring assembly—rear	350 lb/in rate
3927504	Spring assembly—rear	356 lb/in rate
3927507	Spacer—rear spring	Use with HD springs
3889964	"U" bolt	Use with HD springs
SWAY BARS		
3962795	Stabilizer shaft—front	3/4" diameter
3962796	Stabilizer shaft—front	7/8" diameter
3962797	Stabilizer shaft—front	15/16" diameter
3961763	Stabilizer shaft—front	1.0" diameter
3962799	Stabilizer shaft-front	1-1/16" diameter
3927506	Bushing—stabilizer shaft	Use with 1-1/16" bar
3927944	Plate—stabilizer bracket reinf.	Use with 1-1/16" bar

BRAKES (1967 and 1968 RPO J52)
5468882 HD replacement front brake pad unit
. . . . Corvette HD brake pads
5468886 Caliper assembly—front brake, L.H.
. . . . Revised for 5468882 shoe
5468887 Caliper assembly—front brake, R.H.
. . . . Revised for 5468882 shoe
3923535 Pipe assembly—front brake, L.H.
3923536 Pipe assembly—front brake, R.H.
3927510 Hub and disc assembly—front
. . . . 1/2 x 20 wheel bolts
3830635 Shoe and lining unit—rear . . . . Metallic shoes, rear drums
3927508 Shaft assembly—rear axle
. . . . Shot peened 1/2 x 20 wheel bolts

BRAKES (1968 CONVERSION TO FULL DISC BRAKES)
5468882 Replacement front brake pad unit
. . . . Corvette HD brake pads
5452515 Replacement rear brake pad unit
. . . . Corvette HD brake pads
3957992 Front hub and disc w/caliper unit, L.H.
. . . . 1-1/4" disc w/Corvette calipers
3957993 Front hub and disc w/caliper unit, R.H.
. . . . 1-1/4" disc w/Corvette calipers
5463775 Caliper assembly, L.H. . . . . HD Corvette caliper
5463776 Caliper assembly, R.H. . . . . HD Corvette caliper
5463751 Master cylinder assembly . . . . For full disc brakes
3947289 Caliper mounting bracket, L.H. . . . . For Camaro knuckle
3947290 Caliper mounting bracket, R.H. . . . . For Camaro knuckle
3945125 Caliper support brace
3947283 Bracket—brake hose support, L.H.
3947284 Bracket—brake hose support, R.H.
3947037 Brake pipe, L.H. . . . . Connects brake hose to caliper
3947038 Brake pipe, R.H. . . . . Connects brake hose to caliper
3945118 Hub and disc assembly . . . . . Replacement hub and rotor
3945190 Rear disc assembly . . . . Replacement rotor
3945189 Spacer, rear caliper
5457093 Caliper assembly—rear . . . . HD Corvette caliper

REAR AXLE
Special parts for HD disc brake rear axle assembly:
3945131 Rear axle assembly 3.73:1 ratio
. . . . Complete disc brake axle assembly
3941917 Rear axle assembly 3.25:1 ratio
3941918 Rear axle assembly 3.42:1 ratio
3953697 Rear axle housing and spacer unit
. . . . Basic axle housing
3945184 Shaft axle (large diameter)
. . . . For disc brake axle assembly
3945176 Positraction differential case for
3.07-3.73 ratios . . . . HD Positraction—22 plate
3945177 Positraction differential case for
3.91-5.13 ratios . . . . HD Positraction—22 plate
3957937 Gear—differential . . . . Differential side gears
3957938 Pinion—differential . . . . Spider gears
3957939 Spring—differential pinion pin
. . . . Positraction pressure springs
3957940 Washer—differential pinion thrust
. . . . Spider gear thrust washers
3957941 Plate—differential clutch pressure
. . . . Use with springs 3957939
3957942 Unit—differential clutch plate and
disc . . . . 22-plate Positraction repair kit
7451275 Bearing assembly—rear wheel . . . For 3945184 axle shafts
9422406 Bearing assembly—rear wheel . . . . . Optional w/7451275
3843648 Seal assembly—rear wheel bearing
. . . . For large bearings
3851677 Bolt—rear wheel . . . . 7/16 x 20 axle bolts

Standard Camaro part:
3927508 Shaft axle—production drum rear axle
. . . . Shot peened, 1/2 x 20 bolts

RING AND PINION UNITS
3865995 3.55:1 ratio ring and pinion . . . . 12 bolt HD
3865996 3.07:1 ratio ring and pinion . . . . 12 bolt HD
3931564 3.25:1 ratio ring and pinion . . . . 12 bolt HD
3865994 3.31:1 ratio ring and pinion . . . . 12 bolt HD
3931565 3.42:1 ratio ring and pinion . . . . 12 bolt HD
3865997 3.73:1 ratio ring and pinion . . . . 12 bolt HD
3931566 3.91:1 ratio ring and pinion . . . . 12 bolt HD
3917971 4.10:1 ratio ring and pinion . . . . 12 bolt HD
3931567 4.33:1 ratio ring and pinion . . . . 12 bolt HD
3917973 4.56:1 ratio ring and pinion . . . . 12 bolt HD
3917972 4.88:1 ratio ring and pinion . . . . 12 bolt HD
3933095 5.13:1 ratio ring and pinion . . . . 12 bolt HD
3961192 4.kk:1 ratio ring and pinion
. . . . Lower hardness for drag racing
3961195 5.13:1 ratio ring and pinion
. . . . Lower hardness for drag racing

MISCELLANEOUS
9777477 Bolt, wheel hub . . . . 1/2 x 20 wheel bolts
3931547 Wheel assembly (15 x 8" mag) . . . . .2 + offset
3931548 Seat, bucket (plastic) . . . . Lightweight seat
3916633 Rear deck spoiler . . . . Production Z-28, 1968
3949798 Rear deck spoiler . . . . Production Z-28, 1969
3943249 Front valance air deflector
. . . . Large 4" deflector
3943251 Brace valance, outer . . . . For large deflector
3943253 Brace valance, center . . . . For large deflector
3963832 Hood unit, ducted fresh air
. . . . 1969 fresh air—fiberglass
3963827 Stud, hood hold-down
. . . . Fiberglass ducted hood
3963828 Pin, hood hold-down . . . . Fiberglass ducted hood
3963829 Plate, hood hold-down
. . . . Fiberglass hood reinforcement
3963830 Cable, hood hold-down . . . . Pin retainer
3965713 Rod . . . . Fiberglass hood support
3963824 Base plate . . . . Air cleaner base
3963823 Seal . . . . Air cleaner base to hood
3963825 Element . . . . Air filter
3941146 Cover . . . . Air cleaner cover
3016688 Radiator . . . . 2.7" core
3157804 Radiator . . . . Oil cooler

ACCELERATION standing ¼ mile, seconds

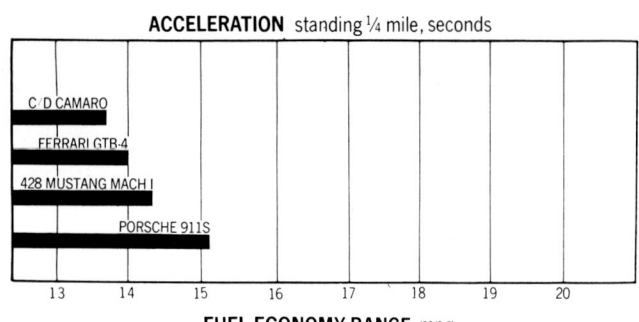

	13	14	15	16	17	18	19	20

- C/D CAMARO
- FERRARI GTB-4
- 428 MUSTANG MACH I
- PORSCHE 911S

BRAKING 80-0 mph panic stop, feet

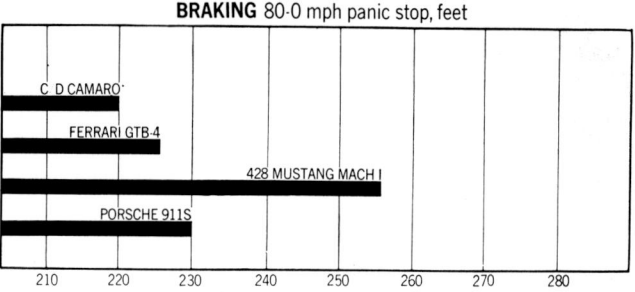

	210	220	230	240	250	260	270	280

- C/D CAMARO
- FERRARI GTB-4
- 428 MUSTANG MACH I
- PORSCHE 911S

FUEL ECONOMY RANGE mpg

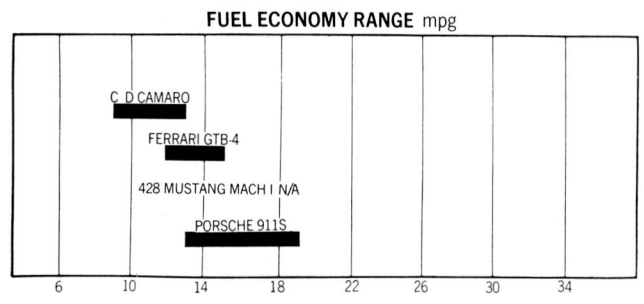

	6	10	14	18	22	26	30	34

- C/D CAMARO
- FERRARI GTB-4
- 428 MUSTANG MACH I N/A
- PORSCHE 911S

PRICE AS TESTED dollars x 1000

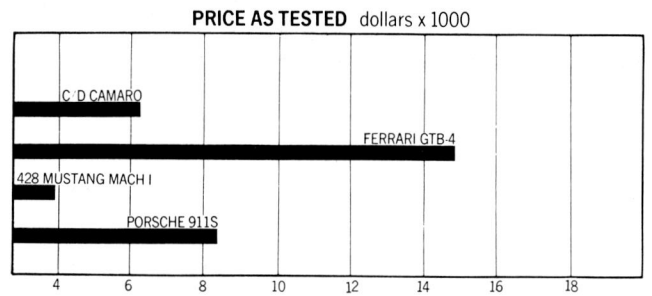

	4	6	8	10	12	14	16	18

- C/D CAMARO
- FERRARI GTB-4
- 428 MUSTANG MACH I
- PORSCHE 911S

C/D CAMARO

Manufacturer: Chevrolet Division
General Motors Corporation
30003 Van Dyke
Warren, Michigan 48090

Vehicle type: Front-engine, rear-wheel-drive, 4-passenger, 2-door coupe

Price as tested: $6221.00
(Manufacturer's suggested retail price, including all options listed below, Federal excise tax, dealer preparation and delivery charges)

Options on test car: 350 cu in LT-1 V-8, N.A.; 4-wheel power assisted disc brakes, $530.30; Stahl headers, $160.00; Koni shock absorbers, $97.00; American Racing 200S wheels, $295.00; Goodyear F60 tires, $304.60; ducted hood, $79.00; 4-speed transmission, $195.40; limited-slip differential, $42.15; variable-ratio power steering, $94.80; adjustable steering column, $45.30; console, $53.75; AM/FM Stereo radio, $239.10; air conditioning, $376.00; tinted glass, $32.65; spoiler, $32.65; shoulder belts, $12.15; Rally Sport and SS equipment, N.A.; custom interior, $110.60; door edge guards, $4.25; bumper guards, $25.30; special instrumentation, $94.80; remote control mirror, $10.55; visor mirror, $3.20; special exhaust system, N.A.; special paint and striping, N.A.

ENGINE

Type: V-8, water-cooled, cast iron block and heads, 5 main bearings

Bore x stroke.4.00 x 3.48 in, 101.6 x 87.4 mm
Displacement.............350 cu in, 5740 cc
Compression ratio.................11.0 to one
Carburetion....................1 x 4-bbl Holley
Valve gear.Pushrod operated overhead valves, mechanical lifters
Power (SAE)...........370 bhp @ 6000 rpm
Torque (SAE).........380 lbs/ft @ 4000 rpm
Specific power output........1.06 bhp/cu in, 64.5 bhp/liter
Max recommended engine speed...6500 rpm

DRIVE TRAIN

Transmission............4-speed, all-synchro
Final drive ratio.................3.73 to one

Gear	Ratio	Mph/1000 rpm	Max. test speed
I	2.20	9.0	56 mph (6200 rpm)
II	1.64	12.1	75 mph (6200 rpm)
III	1.27	15.6	97 mph (6200 rpm)
IV	1.00	19.8	129 mph (6500 rpm)

DIMENSIONS AND CAPACITIES

Wheelbase.............................108.0 in
Track, F/R......................59.6/59.5 in
Length..................................186.0 in
Width.....................................74.0 in
Height....................................51.0 in
Ground clearance........................5.0 in
Curb weight...........................3615 lbs
Weight distribution, F/R........56.2/43.8%
Battery capacity.........12 volts, 61 amp/hr
Alternator capacity..................444 watts
Fuel capacity...........................18.0 gal
Oil capacity.............................4.0 qts
Water capacity........................16.0 qts

SUSPENSION

F: Ind., unequal length control arms, coil springs, anti-sway bar
R: Rigid axle, semi-elliptic leaf springs

STEERING

Type..........Variable-ratio recirculating ball, power assist
Turns lock-to-lock............................2.0
Turning circle curb-to-curb..........40.75 ft

BRAKES

F:......11.75-in dia vented disc, power assist
R:......11.75-in dia vented disc, power assist

WHEELS AND TIRES

Wheel size.........................8.5 x 15-in
Wheel type......American Racing Equipment, cast aluminum
Tire make and size.........Goodyear F60 x 15
Tire type..................Polyglass, tubeless
Test inflation pressures, F/R......26/26 psi
Tire load rating.....1500 lbs per tire @ 32 psi

PERFORMANCE

Zero to	Seconds
30 mph	2.1
40 mph	3.0
50 mph	4.0
60 mph	5.4
70 mph	6.8
80 mph	8.6
90 mph	10.4
100 mph	12.9

Standing ¼-mile......13.7 sec @ 103.68 mph
Top speed (observed)..............129 mph
80-0 mph......................220 ft (0.97 G)
Fuel mileage.......9-13 mpg on premium fuel
Cruising range...................162-234 mi

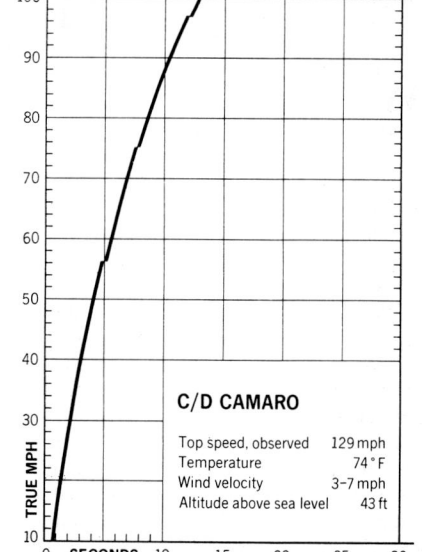

C/D CAMARO

Top speed, observed	129 mph
Temperature	74°F
Wind velocity	3-7 mph
Altitude above sea level	43 ft

Standing ¼-Mile

CAR and DRIVER ROAD TEST

C/D "Blue Maxi" Camaro

We set out to prove that a mere automobile can transcend the obvious device of transportation and become an experience, and we have succeeded

When you set out to transform a Z/28 into a Z/29, the world scoffs, friendly dogs turn on you and bite, relatives shake their heads reprovingly behind your back and the bank reviews your line of credit.

Unless you are a car magazine and you're building The Company Car.

Even then, you are a missionary, guiding your non-existent car through a doubting universe, converting a race driver here, making a believer out of a Firebird 400 owner there. Oh, it is a burden. But not without its rewards.

There are those moments in the early, misty morning when you are alone on a feeder road to a freeway and everything just happens; you push yourself back in the seat with the throttle and for an instant you are sailing down the hill at Laguna Seca or

going through the big bend at Lime Rock. The car comes together, everything you've had put into it, everything it was meant to be, and you go through the corner absolutely perfectly, taking the apex and sliding out under power. In a sad instant the road is over and you are accelerating onto the passionless artery—but even then you find yourself at freeway speeds in seconds, and as you loaf along in the right-hand lane you watch the Plymouth Furys and the Chevy Impalas whirring importantly by and you have an understanding and a memory they will never have.

That is the way it is with the Blue Maxi, and when it first found the bright rays of day shining along its Sunoco Blue flanks, illuminating its fine, fat F60 Goodyears and the tough American Mags which will be next year's hottest wheels, it was appar-

ent we had conceived, ordered and now owned the world's first Z/29 Camaro.

The Company Car: a dazzling, dizzying blue wonder that had no easy goals to meet. It would have to handle with its Trans-Am brothers and do us justice on the drag strip. At the same time it would have to be tractable on the highway and it would have to be a credit to its Publisher. Mostly it had to represent everything we thought was right about the automobile in an atmosphere of increasing criticism of anything with four wheels—especially four *fat* wheels.

Translated, what we expected of Maxi is that it would stop with a decelerative force on the order of 1.0G. It would have to get us through the quarter in under 14 seconds. It would need to be supremely comfortable on long trips, it would have to

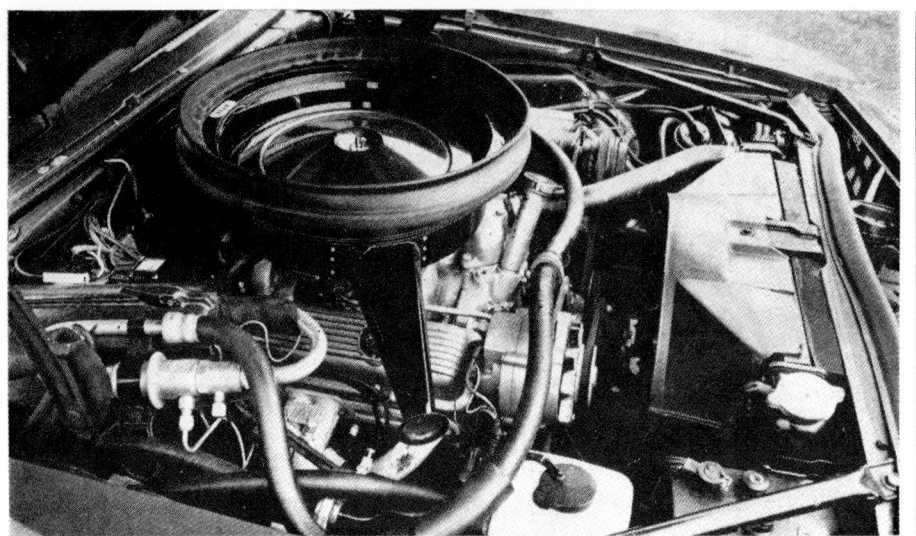

The engine is next year's incarnation of a Z/28, the LT-1, a stroker 302 with 0.48 more arm. It is a simple proposition: What is good small is likely to be better large. That's the whole point in a Z/29

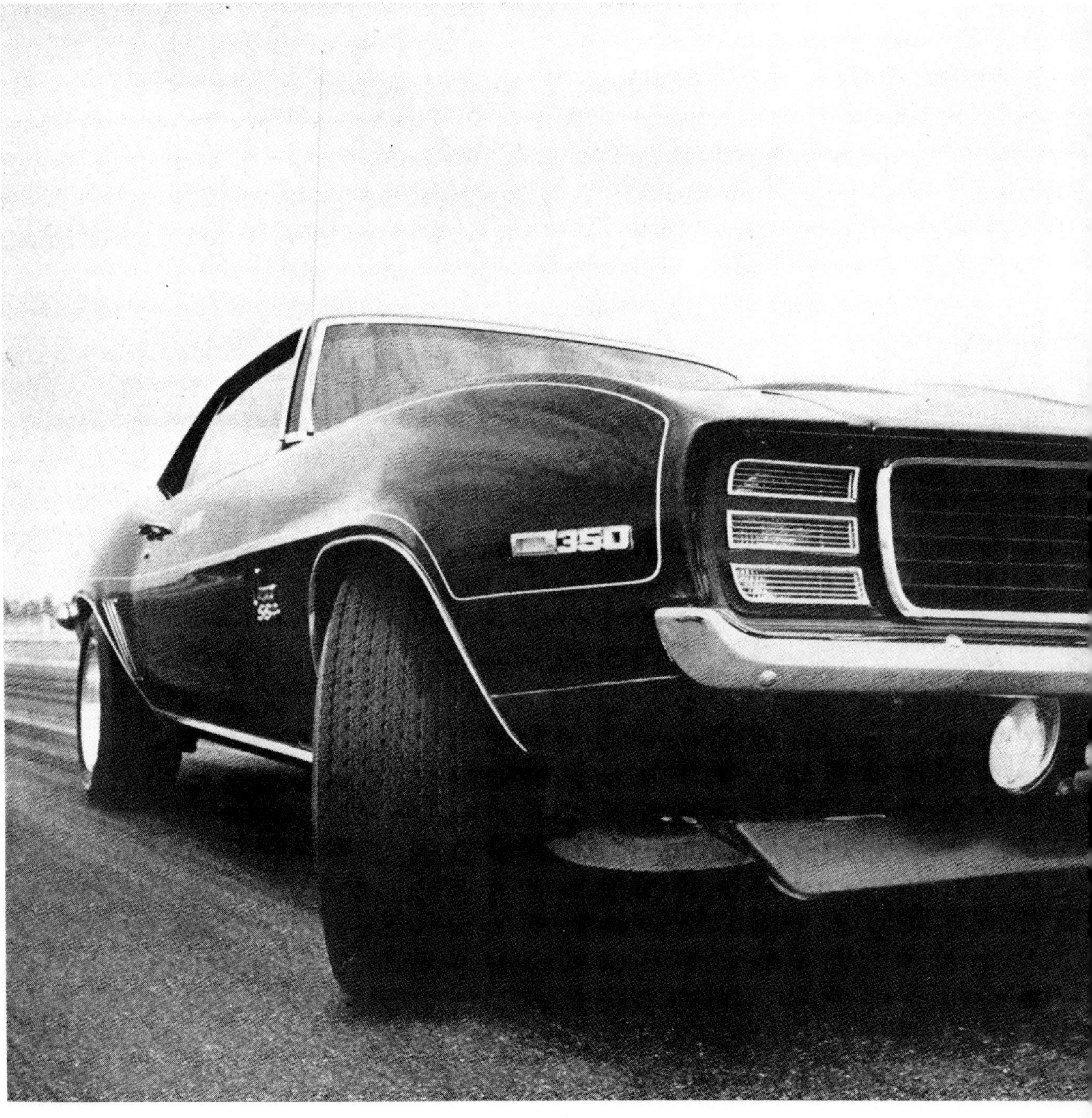

PHOTOGRAPHY: HUMPHREY SUTTON

have a kind of style to it that would make it instantly recognizable and as instantly coveted.

Our choice was a Camaro with the new LT-1 engine (*C/D*, July), and the magician in residence on the project of making a Camaro which was somewhere on the outer boundaries of contemporary practice was Mark Donohue. It is said that Donohue has knowledge of Camaros. His deputy was Sam Eckerd, another of the coven of Roger Penske's warlocks.

Gestation was a matter of three months, and it was an anxious time awaiting birth. And then the car almost miraculously appeared—after untold hours in Philadelphia fitting the engine, the brakes, getting the Stahl headers somewhere inside and finding room for the fat tires.

It was an instant success. With instant problems.

Everyone wanted to drive it, but our Z/29 had a mind of its own. It wasn't satisfied with the way it was set up and began to eat plugs at an alarming rate. Its thirst in no way suffered by having to share gluttony quotient with the engine's spark plug appetite, and very soon there was more Sunoco 260 being pumped in the Greater New York area than had ever been dreamed of by Sun's marketing experts.

But even these seemed minor things compared to what the Maxi was doing to those who did manage to drive it and to the local Chevrolet dealers.

Maybe it was the wheels or the tires, maybe it was the paint job with its vivid striping, but fellow sporty car owners almost drove off the road when they saw The Company Car. At big Byrne Chevrolet in White Plains, New York, where our car went for its first oil change, there was disbelief before the car was allowed even to enter the enormous and sanitary service department.

"A 350 Z/28? You're kidding. There *is* no such thing."

There is no scorn in this world quite like the scorn of a service manager on his own turf and his smirk deepened when he saw the SS emblems on the car.

"Take it around the block once." Just once and the smirk was gone, and that middle-aged gentleman boomed the Blue Maxi into the customers' lane, jumped out and yelled to his performance mechanic to come and have a look. The underground had told him all about the 1970 LT-1 engine and there it was in all its beauty.

Still, there were annoying problems of ride harshness, and the seat was breaking down in an unexpected and unpleasant fashion (a problem the Blue Maxi still suffers) without even giving an acceptable level of lateral support.

Worse still, the car was not quick enough and would not stop when ordered.

Back to Philadelphia for a change in shocks and attention to the proportioning valve for the brakes.

That's all it took. Magic cars aren't sup-

posed to be perfect immediately. You have to live with them and suffer agonies commensurate with the joys they promise. And our car was no exception.

As appears in its final, Z/29 form, the Blue Maxi differs very little from its sibling Z/28. The whole secret was (as we have been telling you for some time) a careful selection of parts—and fine tuning.

The engine is next year's incarnation of a Z/28, the LT-1, a stroker 302 with 0.48 inches more arm. It was actually converted from a 350 cu. in. 350-hp Corvette, vintage 1969. The conversion included the solid-lifter A/28 camshaft, Z/28 intake manifold and carburetor, and 800 cubic-feet-per-minute Holley 4-barrel. Forged 11.0-to-one compression pistons and 4-bolt main bearing caps are a standard part of the package. The result is a 370-hp Z/29 engine compared to the 290 hp of the original 302 unit in the Z/28.

It *is* true that some modifications were made—why else Roger Penske? If there is a pipeline from Chevrolet Engineering, and Chevrolet would deny that *in extremis,* it must make at least a short stop in Philadelphia on its way to Midland, Texas. Initial problems with the carburetion were solved by quiet recalibration by Sam Eckerd, and it would come as no surprise if it were done with the help of coded messages from the factory.

Basically, it was just re-jetted with some additional help in the off-idle transfer system. The timing was also advanced a bit.

The result is a transformation of the engine.

When we first ran the car in preliminary tests the best we could do was 14.2 seconds at 100 mph; with the magic of Penske attendant upon it, the times improved dramatically to 13.7 and the speed to 104.

Equally important, the braking—which was abominable at first—returned figures which are the third best in our history. The car stopped in 220 feet, generating .97G.

The ducted hood is not to be ignored in all this. It is an especially valuable addition on an air conditioned car—and air conditioning was a part of our required package when we decided on what was necessary to produce a thoroughly acceptable GT. With A/C, the underhood temperatures are very high. That not only reduces performance, but also contributes to rough idle and low speed operation.

The air intake system has been modified so that cold air is ducted to the carburetor at all times now—simply by removing the vacuum-controlled valve in the hood scoop and plugging the under-hood air cleaner snorkle. It may have to be changed back for winter time operation.

Since the car itself was originally built as an SS350, the Blue Maxi came with the standard intended-to-be-quiet-rather-than-high performance muffler. It acted as a cork. A later change to a 1968 Z/28 muf-

Text continued on page 86

With a price tag of $4160 on the engine package alone and a tariff of $7800 on the complete car, the ZL-1 Camaro (in showroom shape) leaves an awful lot to be desired. But, once it's breathed on, watch out!

Chevrolet's ZL-1 Camaro: Potential? Yes! Performance? No!

BY ROGER HUNTINGTON

Above, factory built Camaro race car with aluminum ZL-1 engine came through with fresh air hood, spoilers and GT tires. Below, Dick Griffin of drag racing fame tunes Hugger.

THIS IS GOING to be a little different kind of a "road test." When Chevrolet announced that they would build 50 Camaros with all-aluminum ZL-1 Corvette 427 engines to make the combination eligible for Super Stock racing under NHRA rules, many of the car magazines rushed to get their hands on one for strip tests. But it seems like everybody is testing the car after it's all set up for racing, with the engine blueprinted, exhaust headers, big tires, traction arms, etc. This makes sense, as this is strictly a racing car.

But the editors of *Cars* Magazine thought it might be interesting to see what the thing would do just

78

CHEVROLET'S ZL-1 CAMARO: POTENTIAL? YES! PERFORMANCE? NO!

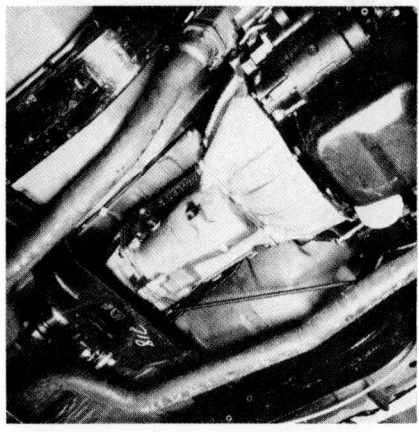

Aluminum-case Muncie four-speed trans weighs less than 70 pounds. Ford four-speed weighs 120.

Aluminum cylinder heads are completely changed. Exhaust ports are larger and have a round shape. Combustion chambers are opened up under the spark plug so they resemble an open hemi chamber.

The all-aluminum block and heads save about 160 pounds compared to the cast-iron 427 version. Total weight of the car with a half tank of gas was same as small block Z-28 car.

a few comments on the car-engine package.

The new aluminum ZL-1 427 engine has been covered on these pages before. Briefly, the engine is somewhat equivalent to the L-88 Corvette racing option, but has a new beefed aluminum block with shrunk-in cast-iron cylinder liners. Also, however, the latest L-88/ZL-1 aluminum cylinder heads are completely changed. Exhaust ports are larger and have a round shape at the gasket face (instead of square), and the combustion chambers are opened up under the spark plug so they more nearly resemble an open hemi chamber. This improves flame travel and reduces shrouding effect around the valves. Breathing is said to be improved 8 to 10 percent when combined with a new camshaft with more exhaust lift and duration. The big-port L-88 aluminum intake manifold and 850-cfm Holley are retained.

In all, the new aluminum block and heads save about 160 pounds as compared with the cast-iron 427 with aluminum manifold. This should put the total engine weight (with accessories, but no flywheel or clutch) to around 520 pounds, or maybe 510. This would be a few

the way it comes off the factory assembly line, with street tires, street exhaust and no more tuning than a check of spark timing. After all, a few well-heeled guys are bound to buy these cars for the street. And it's very possible that a slightly detuned version of this aluminum-engined Camaro might be offered as a super-hot street machine in 1970. Why not?

We got a rare opportunity to check the car out in "showroom" form when Berger Chevrolet in Grand Rapids, Michigan had their ZL-1 Camaro for about a week, before selling it to a private buyer in Virginia. Berger is one of the biggest

high-performance Chev dealers in the Midwest, and they get all the hot models and parts as quickly as anybody. John Grivens, manager of the US-131 drag strip at Martin, Michigan, cooperated by opening his strip for private tests on a weekday. I was also fortunate to have the services of Dick Griffin for the initial tune-up and adjustment and the driving chores on the strip. Dick is one of the better-known Michigan drag racers in the Pure Stock classes—running a Road Runner and 440 Barracuda currently—and he can make a car go in showroom form if anybody can.

Before we get into the testing,

Factory ZL-1 427 engine comes with big Holley 850-cfm 'double-pumper' carb on high-riser aluminum manifold plus wild 356-degree solid cam. Car developed 400 hp at the clutch on tests.

Standard Z-28 exhaust system came on the car —very restrictive with free breathing 427 mill.

pounds lighter than the 302-cubic-inch Z-28 iron engine with aluminum manifold.

I was anxious to get the car on the scales. And, sure enough, the total weight with half a tank of gas was just about the same as a Z-28 Camaro with the 302 engine—or around 3300 pounds. We checked out the front rear weight distribution, and this also matched the Z-28. That's just around 56 percent of the weight on the front wheels (without driver). That's a pretty decent figure for traction and handling. The Z-28 is certainly nice in these departments. And we have to conclude that the new aluminum ZL-1

engine is within a few pounds of the weight of the small-block 302 iron engine.

I know a lot of guys figured the front rear distribution would be close to 50-50 with the aluminum ZL-1 in the Camaro. Not by a jugfull. One shudders to think of the weight balance with a heavy *iron* 427 engine in the Camaro. There would be 58 to 60 percent of the total weight on the front wheels. This would be an impossible traction problem. This is why Chevrolet decided to go with the aluminum ZL-1 for their 50 Camaro Super Stockers. The reason the Corvette has near 50-50 distribution with the 427 engine is that the engine and passenger compartment are set far back, and the independent rear suspension adds weight in the back. On the Camaro the engine sits practically between the front wheels, and the tail of the car is very light.

In most other ways our test car was like a standard Z-28 Camaro. It had the Z-28 heavy-duty suspension front disc brakes, 6-inch wheel rims with E70-15 Goodyear Wide Tread GT tires (with the white lettering). These were *not* belted Polyglas tires, which hurt the traction quite a bit. The new belted tires definitely seem to give the best off-the-line traction of any standard street tires available today, especially in Wide Oval sizes and when inflated to 35-45 pounds. We had a fair amount of rubber on the ground; but the lack of a belt under the tread let it squirm and lose

bite when spinning. Another factor: Camaro fenders are a tight fit around the tires, so you can't even use an F70 section on 15-inch wheels without interference. This limits the rubber without fender mods.

The car came through with 4.10 rear end gears with Positraction, with heavy-duty differential gears and axle shafts. (Standard with big engines in any Chev model.) The transmission was an aluminum-case Muncie close-ratio with factory shifter. But this was the special "M-22" super-duty version—which features beefed synchros, lower gear helix angle and thicker teeth. It's much stronger than the standard Muncie, and is priced at $311 over the standard 3-speed in the Camaro. It's a must on a car like this.

We were a little scared when we looked at the "showroom" exhaust system on the car. It was the standard Z-28 setup, with small resonators ahead of the rear axle and one large cross muffler behind the axle with dual inlets and dual outlets. This isn't too efficient, even with the little 302 engine. We shuddered to think of the loss with the free-breathing ZL-1. Furthermore, there was no room around the engine compartment for the big Corvette streamlined cast-iron exhaust headers. They had to use standard 396 manifolds to fit in. And these didn't even have round openings to match the big round exhaust ports in the heads! We could see another huge loss here.

But, mind you, I'm not condemning Chevrolet for putting this exhaust system on this car. They intend the car strictly for racing. They know the buyer will immediately whip off the complete factory exhaust system and stick on a set of tuned tubing headers. Why worry about spending thousands to develop an efficient factory system? This is a new side of modern Detroit performance engineering. Up to a point the factory boys are concerned with the ultimate in street performance, and they give you their best. But when the model gets so hairy that it's strictly for the racetrack, you'd be surprised how many little details the factory leaves up to you and the California hot rod industry!

But what do you do with this exhaust system to give the car a fighting chance to turn some decent times on the drag strip? Dick Griffin and the Berger mechanics finally solved it by replacing the dual pipe/muffler system with dual "chambered" pipes that are an option on current Che-

Take a good look at it. This may be the closest you'll ever come to an all-aluminum ZL-1 engine. With a price tag of $4100 for the engine package alone and a total cost for a ZL-1 Camaro of $7800 you'd better believe daddy's not going to get you one for your birthday.

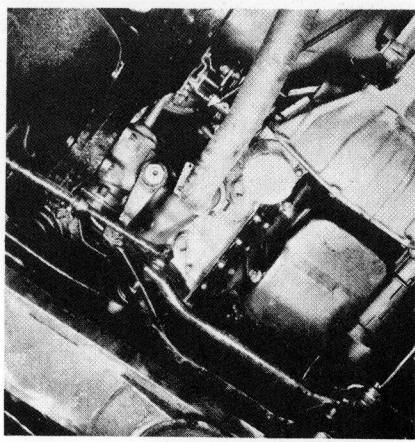

Standard 396 exhaust manifolds have to be installed on 427s because of tight clearance.

velles, Camaros and Corvettes. This is a kind of pipe within a pipe, with holes in the inner pipe and the outer one crimped down every few inches to give a series of resonating chambers along the length of the pipe. The deal is quite a bit louder than a conventional dual muffler system. That's why they stopped using it as standard equipment of the Corvette. But it's a legal street system, it's a factory-installed option on the Camaro—so we thought it would be "fair" to use it for this test. There's no question that it reduced the et several tenths as compared with the factory system.

We hardly knew what performance to expect on the strip. We knew that Corvettes of nearly equal weight will usually turn et's in the mid-13's at 105 to 107 mph with the standard 427 big-port 4-barrel engine (425 hp rated). But we knew that the improved breathing of the new ZL-1 heads should add at least 20 to 30 horses. *If* the exhaust gas could get out through the restrictive manifolds and chambered tailpipe and if the E70 street tires could get a decent bite, we could hope for low 13's and 110 mph.

And that's exactly what we got. The best run was 13.16 et at 110.21 mph. There were six runs between 13.20 and 13.40 and 108-110. Driving technique had Griffin burning lightly out of the hole from 2500 rpm static, feathering the throttle for the first

30 or 40 feet, then on it hard and shifting at 6500 to 6800 rpm. The throttle feathering in low was a bit of a problem. Our test car had the new Z-28 "fresh air" hood, with the air opening facing backward at the rear of the hood. This is supposed to pick up air under light pressure at the base of the windshield. (The way air is forced into the heater plenum chamber under the cowl.) But there is a flap valve in this air opening that is closed by an electric solenoid when the throttle is closed and manifold vacuum goes up (so the engine can cruise on warm underhood air). Needless to say, this deal caused a few problems when feathering off the line. The flap valve fluttered open and shut and kept the engine from getting a good breath. When this valve was wired full open, the car came out of the hole better.

That's most of the story, gang. This is what you could expect from a showroom ZL-1 Camaro on the street, with legal street exhaust and street tires. (Though you could do better in both departments with some special stuff.) The ZL-1 camshaft was definitely smooth enough to drive on the street, at least in warm weather, though the idle was quite rough. Young performance fans are learning to live with wilder and wilder cams these days, so this engine should cause no special headaches. One sober note was that the rich-jetted 850-cfm Holley carb gave only

5 to 7 mpg in normal street driving! But when you jump on that loud pedal, all is forgiven!

Incidentally, those new ported-and-chambered ZL-1 heads *did* add some horses to the engine. My slide rule says it would take around 400 honest horses at the clutch to accelerate the 3300-pound Camaro to 110 mph in a quarter-mile. I figure the average showroom 427 Corvette engine with the big-port heads develops about 380 hp at the clutch, without any special tuning. And this is with the streamlined cast-iron exhaust headers. This ZL-1 pulled 20 horses more with the regular 396 manifolds. That's pretty good when you come right down to it. It should be no trouble getting 600 to 650 hp in all-out Super/Stock form.

Which class? NHRA officials have just quoted a pounds-per-hp factor of 6.50 for the car—which puts it right in the top of SS/C. They "factored" the hp rating from 430 to 480, so this would permit you to cut the racing weight of the car to 3120 lbs. Prototypes have already tested out with et's in the low 10's at 130 mph with blueprinted and modified ZL-1 engines. That will clobber the SS/C class. Don't forget that the hot Hemi Barracudas and Darts are factored to 525 hp, and have to run in the SS/B class. If these ZL-1 Camaros are given a couple of tenths handicap start on them in Super/Stock bracket eliminations, it could be all over for the Mopars. Chrysler fans should do some screaming.

But, then again, maybe there won't be enough ZL-1 Camaros around to notice. Chevy puts a price of $4160 on the engine package alone! The car we tested from Berger had a sticker of *$7800* on the window! Anybody for ZL-1 Camaros for the masses?

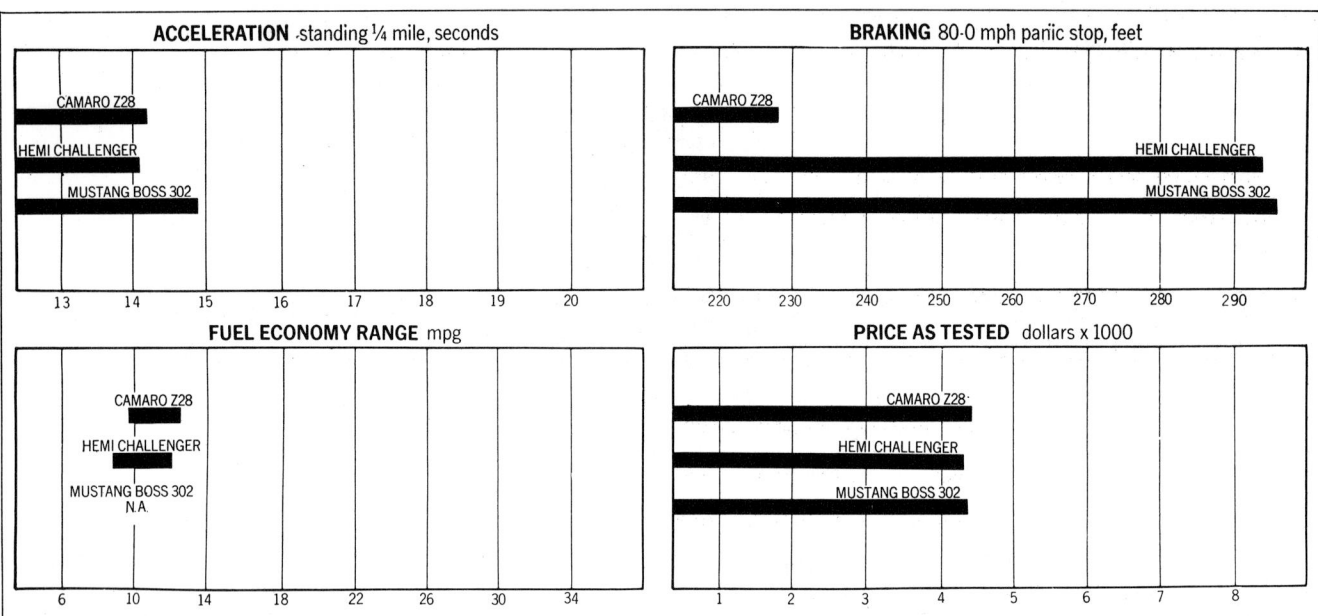

ACCELERATION -standing ¼ mile, seconds

CAMARO Z28	
HEMI CHALLENGER	
MUSTANG BOSS 302	

13 14 15 16 17 18 19 20

BRAKING 80-0 mph panic stop, feet

CAMARO Z28	
HEMI CHALLENGER	
MUSTANG BOSS 302	

220 230 240 250 260 270 280 290

FUEL ECONOMY RANGE mpg

CAMARO Z28	
HEMI CHALLENGER	
MUSTANG BOSS 302 N.A.	

6 10 14 18 22 26 30 34

PRICE AS TESTED dollars x 1000

CAMARO Z28	
HEMI CHALLENGER	
MUSTANG BOSS 302	

1 2 3 4 5 6 7 8

Camaro Z/28

Manufacturer: Chevrolet Motor Division
General Motors Corporation
Detroit, Mich. 48202

Vehicle type: front engine, rear-wheel-drive, 4-passenger coupe

Price as tested: $4475.70
(Manufacturer's suggested retail price, including all options listed below, Federal excise tax, dealer preparation and delivery charges, does not include state and local taxes, license or freight charges)

Options on test car: Base V-8 Camaro, $2839; rally sport, $168.55; custom interior, $115.90; door edge guards, $5.30; visor mirror, $3.20; Z/28 package, $572.95; automatic trans, $221.80; limited slip diff., $44.25; power steering, $105.35; power brakes, $47.40; custom seat belts, $12.15; console, $59.00; tinted glass, $37.95; special instrumentation, $84.30; auxiliary lighting, $11.10; sport mirrors, $26.35; AM radio, $61.10; rear speaker, $14.75; tilt wheel, $45.30

ENGINE
Type: V-8, water-cooled, cast iron block and heads, 5 main bearings
Bore x stroke 4.00 x 3.48 in, 101.6 x 88.4 mm
Displacement............350 cu in, 5740 cc
Compression ratio..............11.0 to one
Carburetion..............1 x 4-bbl Holley
Valve gear........pushrod operated overhead valves, mechanical lifters

Power (SAE)...........360 bhp @ 5600 rpm
Torque (SAE)..........370 lbs-ft @ 4000 rpm
Specific power output....1.03 bhp/cu in, 62.9 bhp/liter
Max recommended engine speed...6500 rpm

DRIVE TRAIN
Transmission............3-speed, automatic
Max. torque converter..................2.1
Final drive ratio..............4.10 to one

Gear	Ratio	Mph/1000 rpm	Max. test speed
I	2.48	7.4	48 mph (6500 rpm)
II	1.48	12.3	80 mph (6500 rpm)
III	1.00	18.2	118 mph (6500 rpm)

DIMENSIONS AND CAPACITIES
Wheelbase.........................108.0 in
Track, F/R....................61.3/60.0 in
Length............................188.0 in
Width.............................74.4 in
Height............................50.1 in
Ground clearance...................4.5 in
Curb weight......................3640 lbs
Weight distribution, F/R........56.7/43.3%
Battery capacity..........12 volts, 61 amp/hr
Alternator capacity................444 watts
Fuel capacity....................20.5 gal
Oil capacity......................4.0 qts
Water capacity...................16.0 qts

SUSPENSION
F: Ind., unequal length control arms, coil springs, anti-sway bar
R: Ind., rigid axle, semi-elliptic leaf-springs, anti-sway bar

STEERING
Type.........Variable-ratio recirculating ball, power assist
Turns lock-to-lock.......................2.6
Turning circle curb-to-curb............37.5 ft

BRAKES
F:.........11.0-in vented disc, power assist
R:..........9.5 x 2.0-in drum, power assist

WHEELS AND TIRES
Wheel size......................15 x 7.0-in
Wheel type.......styled, stamped steel, 5-bolt
Tire make and size..........Goodyear F60-15
Tire type..........Fiberglass belted, tubeless
Test inflation pressures, F/R.......24/24 psi
Tire load rating.....1500 lbs per tire @ 32 psi

PERFORMANCE
Zero to	Seconds
30 mph	2.1
40 mph	3.1
50 mph	4.4
60 mph	5.8
70 mph	7.3
80 mph	9.2
90 mph	11.6
100 mph	14.2

Standing ¼-mile..........14.2 sec @ 100.3 mph
Top speed (observed)..............118 mph
80-0 mph..................228 ft (0.93 G)
Fuel mileage...9.5–12.5 mpg on premium fuel
Cruising range....................195–265 mi

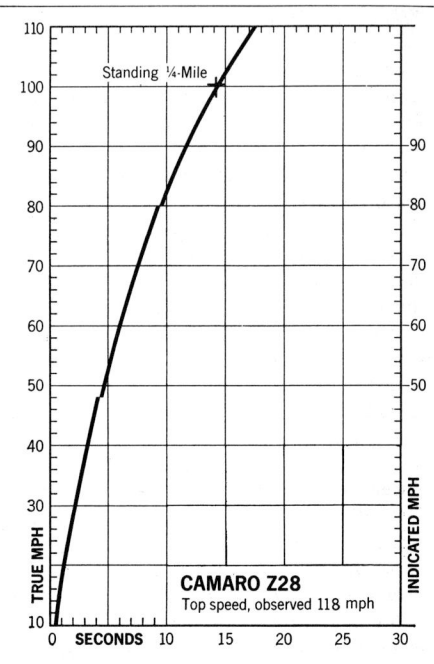

CAMARO Z28
Top speed, observed 118 mph

Standing ¼-Mile

CHEVROLET CAMARO

THE Z/28 VERSION WOULD BE EVERY BIT AS MUCH AT HOME ON THE NARROW, TWISTING STREETS OF MONTE CARLO AS IT IS ON INTERSTATE 80

The world's menu of powerful GT cars contains a few selections of uncommon merit. Almost invariably they are European, frequently Italian in descent, few in numbers and high in price—the precious gems of the car builder's art. There is nothing precious about the Camaro Z/28, Chevrolet will stamp them out like the government does cupro-nickel quarters, but it is an automobile of uncommon merit. It would be every bit as much at home on the narrow, twisting streets of Monte Carlo or in the courtyard of a villa overlooking the Mediterranean as it is on Interstate 80. It's a Camaro like none before.

As everyone knows by this time the 1970 Firebird and Camaro were introduced late in February rather than during Detroit's annual fall festivities. Several weeks before Camaros were due in the showrooms, Chevrolet turned loose a half-dozen Z/28s to various members of the automotive press for whatever kind of evaluation pressmen make. We had requested one with an automatic transmission

—the Blue Maxi, with its 350 cu. in. Z/28, had already convinced us that the manual transmission combination was more than satisfactory—and such a car was waiting. It was an early production model that had been carefully inspected and the result was an automobile of commendably high quality.

Almost all car flakes dream of driving some new car before it's available to the public and it can be a completely unique experience. It can also give you an insight not otherwise possible. One young man summed it up best. "I know it's a Z/28, but a Z/28 *what*?" Only the hard core car underworld knew that it was a Camaro. The rest had to ask. And although we think it's a stunning machine from almost any vantage point, it generally went unnoticed—even in Los Angeles where the car reigns supreme. It's a hard situation to explain. We can only theorize that the Camaro's finely drawn shape, free of Detroit's customary visual trickery, is somewhat removed from the mainstream of public taste. Indeed, if the world approves

of Monte Carlos and Rivieras, the Camaro must be an eyesore. We can only hope that is not the case. In fact, there is good reason to believe that the Camaro and Firebird are the leading edge of a new trend in Detroit styling. The Europeans, particularly Giugiaro, have popularized the concept of a strong, simple shape with extreme tumblehome and tuckunder that doesn't need stick-on ornamentation to make it work. The Camaro is certainly of this school. Only the high, pointy grille seems inconsistent with the rest of the car.

And as the styling is restrained in comparison to past Camaros so seems to be the performance image. The cold air induction hood is gone now and so is the Z/28's front spoiler. A change in the Trans-Am rules allows the racers to use a front spoiler whether or not one is available from the factory, and Chevrolet figured it was no longer worth the trouble to bolt them on at the production line since customers just knocked them off on curbs and snow banks anyway. But the cold air hood is another matter and the racers *need* that. Actually, it's not so much performance as performance image that's been dulled. The engines, which have most of the say about performance, are stronger than ever. The Z/28 is richer by 48 cubic inches and 70 rated horsepower (tough break for those with insurance worries), 350- and 375-hp 396s (now actually 402 cubic inches) are still on the list and a 454 lies hidden in the fine print. No discrimination against thrill seekers there.

Somehow, though, the Z/28 is not as thrilling as it once was. It's more tolerant to driving techniques now, more mature in its behavior. All things considered, it's a better engine now but the loss of a carefree and irrepressible adolescent spirit can never be witnessed without some regret. And although the Z/28 seems much tamer now than it once did, the transformation is more a function of the car than of the increase in displacement. The mechanical lifter valve gear still makes its busy clatter and the exhaust pulses still cascade and reverberate through the pipes

THE NEW Z/28 IS
MORE TOLERANT THAN
PREVIOUS VERSIONS
AND EVEN IF IT IS
IRRATIONAL ONE
CAN'T HELP BUT REGRET
THE LOSS OF ITS
ONCE CAREFREE AND
IRREPRESSIBLE
ADOLESCENT SPIRIT

with the same abandon they always did, but the sound engineers have so diligently sealed off the passenger compartment that all of those endearing vibrations are filtered out somewhere before they reach your ears. It's a whole different kind of car now. Better, but not unilaterally. In their zeal, those persecuters of noise have even gone so far as to clamp a silenced air cleaner down on top of the Holley 4-bbl. (automatic transmission versions only). The result is a car of brilliant performance for its displacement and with prep school manners—*not* the combination that brought the Z/28 to pre-eminence within the car culture.

But it does deliver the performance. The automatic transmission test car had a little help from a 4.10-to-one axle ratio, not exactly what we would have dialed up of our own free will. The surprising part is that the high-winding gear is relatively tolerable in this speed-limited United States, provided you're not short of gas money. The interior noise level is so modest that, had we not known about the ratio, we might have suspected some afflic-

tion of the tachometer. If you're interested in acceleration, however, the 4.10 is no more than strategic excess. The Z/28, despite its increase in muscle, is still soft on the low end and with the automatic it would probably bog with a lesser gear. The test car suffered no such infirmities: 14.2 seconds at 100.3 mph in the quarter is as good a measure as any of its physical fitness. Because of the high coolant temperatures required for emission control the power drops off as the engine reaches operating temperature. When fully warmed up the Z/28 is 1-2 mph slower—a situation avoidable when outside air is ducted into the carburetor.

Predictably, the automatic is a great pain reliever when you're beset with a traffic jam. And with the console shifter you are better off leaving the shifting to whatever makes it automatic. If you try to do it yourself you will probably, unless you sandpaper your fingers, lose your way through the shifting maze. Because the detent for the 1-2 shift is indiscernible the result is too often 1-3. Chevrolet promised a motorcycle-type ratcheting shifter several years ago but as soon as our backs were turned it reneged.

Or perhaps the engineers became involved in something else, steering and suspension for example. The new Camaro has completely redesigned steering linkage, now located forward of a line through the ball joints rather than behind as in most other cars. To reduce noise and ride harshness all suspensions have a certain compliance, or ability to deflect, built in. With the linkage mounted forward, the compliance toes the wheels in an understeering direction which contributes to more manageable transient handling. Along with this, all power-steering Z/28 and SS Camaros have a special high-effort steering gear. High effort is not to be confused with increased road feel but it does reduce the tendency to overcorrect. The result of these two developments is a car of exceptional road handling—probably the best Detroit has ever produced. The transition

as you enter a curve or change is extremely predictable and this, combined with a low body roll angle, is the essence of good *road* handling. In more demanding situations, those which you would encounter on a race track or perhaps on a road you had all to yourself, the Camaro is disappointing. It understeers heavily; sometimes you can trick it and get the tail out, sometimes you just have to slow down until the front tires regain their hold on the pavement. Never does it offer the driver very much road feel and never does it give him any confidence. The engineers admit being faced with a compromise, ultimate cornering ability or transients, and they chose the latter. And the driver's lot is made even more difficult by the seat which is just the inverse of a bucket—a

seat that is easier to fall out of than in to.

Another thorn which the engineers had to consider was the limited-slip differential which increases understeer in direct proportion to its limited-slip qualities. Since it's not standard equipment—and oversteer is considered to be dangerous—the base car must have built-in understeer. It follows, then, that the limited-slip car will have even more understeer. The logical solution would be to make the limited-slip a mandatory option since, on a performance car, most buyers order it anyhow and then tune the front and rear anti-sway bar rates for that situation. But for now you'll have to take the Camaro for what it is, a highly developed touring car, and don't expect too much in more demanding situations.

Certainly the brakes are up to any touring demands. Front discs are standard and the test car had the optional power assist. This car represents something new for Chevrolet in that high pedal pressure was required to produce impending lock-up rather than the normal touch on the pedal. The stopping performance was very good, slowing from 80 to 0 mph in 228 feet (0.93G). Directional stability was extremely good because the front wheels locked up first and, while fade was made apparent by an increase in pedal pressure, the three test stops were nearly identical in length.

The mechanical Camaro is obviously successful in its performance and the engineers responsible for that have also come up with a few more subtle technical innovations. One is the Delco battery which secures its cables to the posts with threaded fasteners rather than the traditional clamping method that not infrequently works loose or corrodes into ineffectiveness. Another is the styled wheels which, while they look very much like cast alloy wheels, are actually welded together from a conventional steel rim and a deeply drawn steel center. Although they are no doubt heavier than the standard wheels, and far heavier than the real magnesium wheels they imitate, they should not be subject to the normal problems (corrosion, low impact strength and lug nuts working loose) that plague the general run of cast wheels. And the last bit of technical wizardry that caught our eye is the glove box door hinge which has no moving parts. It's a strip of plastic—one side bonded to the dash, the other to the door—and the strip bends when the door is opened. It's one hinge that will never need oiling.

One of the most conspicuous features of the Camaro's layout. and not necessarily an improvement, is the new long doors. No more is there a rear quarter window, it's all in the door. And not only are the doors longer but they are also moved rearward in the body. Entrance to the rear seat is decidedly easier, but to the front is harder. In a narrow parking slot the exit space for a front seat passenger is inconveniently small and the process is made more difficult by the door's excessive weight, partly attributable to its length and partly to the side impact beams enclosed within.

Of course, the elimination of a window on each side should reduce the chance of wind noise but no such luck on the test car. Chevrolet is trying a new type of window seal on the Camaro and the assembly line workers obviously haven't figured out how to fit it on the car yet. The result was a chorus of wind whistles—the only objectionable noises to be heard in an otherwise sound quarantined car.

The spirit of compromise so apparent in the Camaro's handling and door arrangement carries over strongly into the interior design. The dash now groups all of the instruments directly in front of the driver— a fantastic improvement from the optional gauge cluster that looked up at you from the console in past Camaros—but the small gauges now are *very* small and no matter what your height, some of them are likely to be blocked by the steering wheel. And while we are on the subject, the optional gauges of the past are still optional. Only the speedometer and fuel gauge are standard.

Another of the more obvious compromises is the optional console. It sticks up somewhat higher than the seat cushions and has two recesses in each side in which to stow seat belt buckles. The problem is that the recesses. which are not very handy

for their intended purpose, tunnel so deeply into the console that the bin inside isn't big enough even for road maps. There is another bin, this one open, where the console joins up with the lower edge of the instrument panel. Its utility is not what it could be, primarily because the shift lever, when in park, blocks the opening.

Generally, the Camaro's interior is quite hospitable. Visibility forward is very good because of the narrow, curving windshield pillars, and the wide C-pillars to the rear are less obstructive than they would appear from the outside. The seats in the test car were more upright than those of the early press preview cars (C/D, March) and the driving position suffered slightly. We particularly like the Camaro's inner door panels which are molded of a soft material that gives the sensation of deep padding. It is far more appealing than the hard panels used on Chrysler's sporty cars.

The Camaro, like all cars from Detroit, is a series of compromises, one upon another. At least in the Camaro they've all been made in pretty much the same direction, that of a stylish, quick grand touring car, and the final combination is well suited to its task. And yet, even though the Z/28 is not at all race car-like, some of the strongest suggestions of its competition potential are right on the surface. Those to whom power bulges and love mounds are the only readable evidences probably wouldn't notice, but check the way the glass is nearly flush with its surrounding sheetmetal and the absence of drip rails over the side windows. That wasn't done for gas mileage. After two years of being Trans-Am champion things are expected of Camaro, and if John Delorean and Jim Hall both like this one it has to have something going for it besides nice manners and a pretty fender. ●

C/D "BLUE MAXI"

(*Continued from page 77*)

fler not only helped the sound but added about 2 mph to the trap speed in the quarter. A free-breathing engine like the LT-1 can't stand a restrictive exhaust.

Air conditioning adds 97 pounds to the weight of the car, 89 of them on the front wheels. This is just enough to require chassis modifications if the goal of fine balance in handling and braking is to be achieved.

The initial problem in braking was that the rear wheels unloaded too much during hard application, with resulting rear wheel lock-up and axle tramp. A proportioning valve compensation solved that nicely.

At present, the car is an understeerer, though not unpleasantly so. To overcome it requires only 4 psi more in the front tires, but if it is decided that it is a really bothersome characteristic, the rear springs would have to be less stiff than the dealer-available Camaro HD units yet stiffer than the standard springs.

We have lived with our Z/29 for three

months and it is like living with a beautiful woman. She's there, she's yours, she's unpredictable. You marvel the magic will last. It is all astonishing and not for mere mortals.

The Blue Maxi has served as the course car at Bridgehampton during a driver's school where flat-out speed and evasive maneuverability spell the difference between survival and statistic. We were mobbed by people who wanted just one lap in the car. Those who got it wandered away in shock. A "road car" that was more of a race car than most of them had ever been in, no less driven.

Driven like a race car, the Z/29 has some very endearing characteristics. It is easy to drive. We said last year (C/D, July, 1968) that we wondered, along with test driver Sam Posey, why power steering did not find more currency in racing equipment. Maxi emphasizes the point. The car is enormously responsive, easy to direct

and is almost flawless in transient situations.

As delightful as it is to find a race driver who gets in our Company Car and wails around the track to become an instant convert, it is, if anything, even more delectable to feel the car on the street and bask in the attention it draws.

For most people, Blue Maxi is an *event*, a moment's surcease in a drab and unrewarding day. Properly so. We set out to prove that a mere automobile can transcend the obvious device of transportation and become an experience, and we have succeeded.

C/D's car will find its way into the hands of people who know about cars in the months to come. Not very many, just a few who understand. And what they will understand when they drive it is that until that moment they haven't understood at all.

That's what the Blue Maxi does for you. And that was the whole idea. ●

PROFESSIONALS AT WORK

Watch the Penske Team build a Trans-Am winner, step by legal, painstaking step.

BY DAVID BEAN, ENGINEERING EDITOR

THE WICKED FLEE when no man pursueth. Roger Penske does not flee when pursued. Before the West Coast Trans-Am races, the finish of the 1969 season was clear. Penske's Sunoco Camaro team would win.

We must find out how Penske and his crew do it, but how? We don't know any spies and we can't afford bribery. Besides, we'd probably get punched in the nose, and we are writers, not fighters.

Start at the top, with Penske himself. We wanna come look at your cars. Sure, he said. Here's where we'll be working on them between races. Drop in any time, with notebook, camera, sketchpad, and tape measure. We have no secrets and we can't afford to cheat.

Accusations are easy to make, and hints and snide remarks fell like rain during the 1969 season. There were some flagrant examples of rule break-

age and bending. But Penske is right. He can't afford to cheat. There are too many eyes on him, and he would lose more than one race if he was ever caught.

It's worth noting here that Penske has always been one step ahead of the field when it comes to making the book work for the car. He's been in racing, as a title-winning driver before he became an entrant, for 10 years. He probably holds the record for being protested. And not one protest has ever been upheld. Not one.

Building a Trans-Am winner starts at the factory. This is not to say the Sunoco team is a factory team. If actual money changes hands, nobody has ever found the pipeline, and it's not for lack of effort, inside General Motors and out. The only fair conclusion is that Penske and the factory trade information and advice, just as Penske and the factory say they do.

The results of this interchange show

up in the homologation papers, the list of parts and specifications that determine what can be used on a Trans-Am car and what can't. The manufacturer must do this for the entrant, and Chevrolet does it, for Penske and anybody else who wants to go racing with his Camaro.

But unlike Ford, which is willing to design, manufacture and homologate anything that could possibly be needed for racing, Chevrolet engineering must be able to assign the design cost to some finite account. There is no racing budget at Chevy, so most of the special parts wind up as "Heavy Duty" and work their way into real honest production. In other words, if you were smart enough, you, too, could build a Camaro as fast as Mark Donohue's, while on the other hand, just to get the right parts out of Ford you'd have to at least be named Carrol Shelby.

After making sure all the requisite

PROFESSIONALS

continued

options are registered, the next thing is to establish a technical rapport with the factory. Here Penske made a very wise move early in his association with Chevy. With every enthusiast engineer in the house wanting to get involved in the racing project, and their enthusiastic bosses more than willing to let them, Penske put his foot down. He had seen the factory people overrun the Shelby works and would have none of it in his team. He would like only one engineer, thank you, a liaison man to coordinate the technical info passing in both directions. This kept the operation small, efficient and extremely fluid, yet when major technical assistance was needed it was available.

Penske preferred using his own people. Something that isn't commonly known is that one of the better racing engineers in the country is Mark Donohue. He is not only the Penske team number one driver, he is also the chief engineer. He makes the final decisions about various suspension changes, he does much of the design work, and all the test driving. If any one man can be called responsible for the winning design it has to be Donohue. He is ably backed up by Chuck Cantwell, former Shelby Racing team manager, much of whose time is spent on the drawing board between races. He, too, handles much of the conceptual design, supervises the manufacture of the special parts, and takes care of the complicated logistics of operating a racing team and supplying parts clear across the country.

Each car is cared for by a chief mechanic and a helper. Earl McMillan, a long time Penske mechanic, handles the Donohue car, while Woody Woodward takes care of Bucknum's mount. A pretty small operation, about one quarter the size of any opposing them. About the only work handled by the factory is the stuff the team has no facilities for, such as computer time and elaborate testing programs.

The Penske Camaros start life as nothing more than a pile of "white" parts—components without paint. The factory has nothing to do with chassis construction. Penske gets the parts through the usual channels of his Chevrolet dealership. The unit-body section is lightened where possible (we could find no evidence of acid dipping—usually detectable by the deterioration of spot welds), and the roll cage is welded in, becoming as much spaceframe as roll-over structure. The front subframe is then bolted solidly to the unit-body (rules prohibit welding it) eliminating the rubber noise-suppression stock mounts. The subframe is all rewelded, the roll cage is extended to meet the forward rails and all unnecessary brackets and bolts are

PRESSURE connections on both the radiator and crankcase allow water and oil to be added to running engine during frantic pit stops. Pressurized cannister is snapped on this receptacle, air pressure injects fluid into engine.

FUNCTIONAL instrument panel, cool air vent and foot rest aid driver efficiency. Note oil injection connection behind wheel spoke. This allows oil to be added without lifting hood. Gray paint gives clean appearance, makes leaks readily visible.

eliminated. The front fenders and grille assembly are also reinforced, unnecessary metal is removed, and the panels are arranged so the whole assembly attaches by four bolts. This facilitates easy removal for accessibility to the major mechanical components.

Probably the biggest secret of any Trans-Am racer is the way the suspension works. Standard Ponycars have poor geometry, and even worse torque control. They were meant to be cheap first, sophisticated second. It is a tribute indeed that the racing teams have made sophisticated racing cars out of what basically are inferior designs. (Most full-sized sedans have vastly superior suspensions.)

The major change is the camber change of the front wheels. Standard suspension allows the wheels to lean out during cornering roll, which reduces the cornering power. And since rules prohibit relocation of the suspension pivot points, the improvement must come from subtle adjustment within the existing design. A favorite trick last year amongst all the teams was to lower the inside upper A-arm pivot point. SCCA got wise to this though and now measures it at the tech inspections. For the '69 Camaros, a slightly longer spindle was homologated, effectively accomplishing the

same thing without breaking any rules.

Next, eccentric bushings of Devlon (a tough plastic material) replace the usual rubber bushings, and are used to lower the upper A-arm pivot, and raise the lower. The entire car is also lowered, giving the lower A-arm a near stock angle, while pulling the upper arm into a much steeper angle. The resulting geometry is much better but the extreme static negative camber used (3°) indicates that it is still far from ideal. Still within the normal alignment adjustment, a small amount of anti-dive is used, by tipping the upper A-arm axis slightly towards the rear. Chevrolet has various rates of front spring coils available on a special parts list and these are changed to suit the specific course. There are also several front anti-roll bars cataloged, but the Penske cars now use a custom-made version, with the lever arm length adjustable for tailoring the roll rate. This is simpler and faster than having to swap the entire bar, as would be required with the factory parts.

Power steering was tried early in the season, and although it reduced driver fatigue, it was one more part subject to failure (which it did). When the chips went down early in the season, they decided to eliminate all undeveloped ideas to concentrate on win-

ning. The optional quick 17:1 steering (RPO N44) is used, and with the big tires, it is horrendously stiff. At speed, it lightens, but drivers *must* stay in good physical shape to drive the two- to three-hour races and avoid accident-causing fatigue. Only trick done to the steering is one subtle geometry modification. A small spacer is placed between the steering arm and the tie rod, to lower the tie rod end, causing its arc during suspension travel to be compatible with that of the A-arms, eliminating bump and roll steer.

As the owner of any hot Ponycar will tell you, the worst flaw on these chassis is the rear axle location. The classic Hotchkiss drive is standard fare on all Ponycars, using only the leaf spring to control torque reaction. Spring wrap-up under braking and acceleration is common. Making the spring stiffer helps, but the ride suffers. Torque bars will eliminate wrap-up, but induce a bind, no matter how carefully the pivot point is selected. Challenging problem.

Ford sidestepped it by using two radius rods on each side and taking the locating function away from the spring which sits loosely in slotted mounts.

The Camaros have a more interesting solution. They use no control arms of any kind. The springs are built

LUG nut holder allows all five nuts to be started at once. Air wrench tightens them down. Tire changes are in the 20-second category, fast by even NASCAR standards.

PHOTOS BY GENE PALMER

FRONT SUSPENSION presents formidable challenge on the Camaro. Production geometry is poor, and SCCA don't allow no messing 'round, 'round here. By using a slightly longer spindle, eccentric plastic bushing (in place of the rubber ones) and all the available alignment adjustment, some facsimile of decent geometry can be achieved. Corvette brakes have Teflon coated aluminum hub to ease wheel installation. Adjustable antiroll bar replaces standard unit.

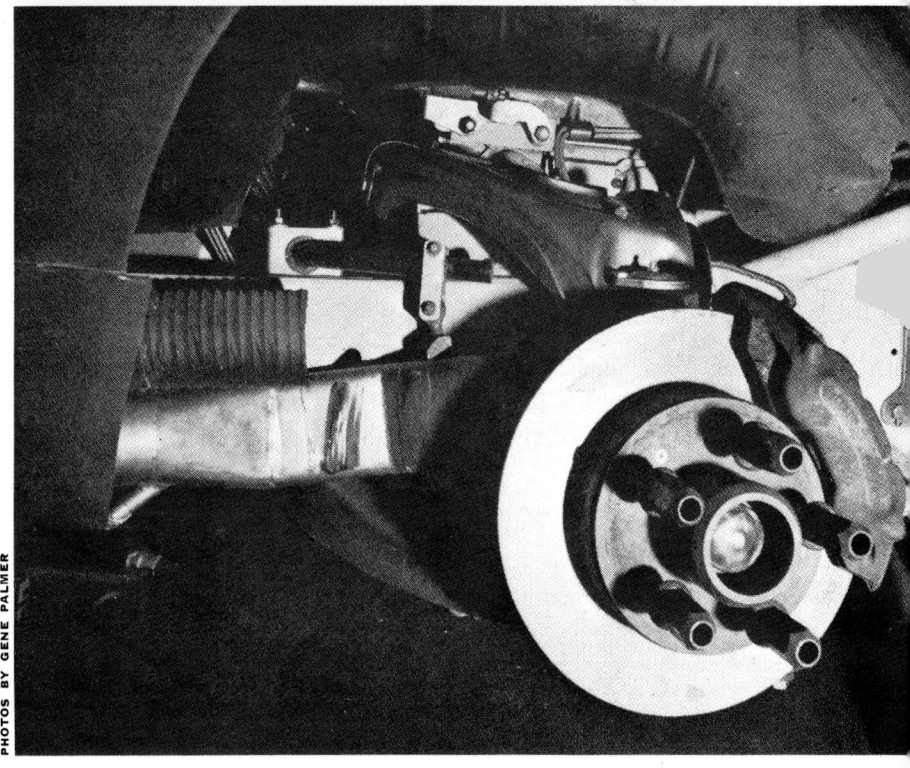

PROFESSIONALS

continued

eccentric, that is, the forward portion is much stiffer than the rear, so that the forward portion acts more as a link than a spring. The rear portion supplies most of the springing. This affords good axle control without having to go super stiff in spring rate.

As exhibited on the race course they are noticeably softer than the competition. These springs also have a very long effective radius of rotation designed in, i.e., the effective center is much further forward than normal, so that during suspension travel, the axle describes a very flat arc, reducing bump and roll steer to a minimum.

Lateral location is provided by a track or Panhard bar. Although a centrally located Watt link achieves straight line travel, it also gives the axle a decidedly unnatural geometry. Suppose a bump is encountered by

one wheel. The axle's natural tendency is to ride over the bump by rotating about the opposite wheel. If the axle is constrained in the center, the axle will rotate about this pivot point causing a like disturbance at the opposite wheel (a slight inward scrubbing movement). With a Panhard bar, the axle is allowed to move in a more natural fashion during one wheel bump, but does cause a lateral movement when both wheels bump. It's a trade off. Stability in a corner is more desirable than stability on the straights.

Placement of the Panhard bar is also a very ticklish proposition. Located wrong, it can cause a jacking action that would terrorize even the most jaded owner of an early Corvair. Jacking isn't the word for it: More like pole vaulting. Obviously more than the drawing board time of Mark

Donohue went into the design of these springs and Panhard bar. It was the result of much computer time.

Devlon bushes are also used at the rear springs, and a half inch adjustable anti-roll bar is mounted above the axle. Koni adjustable shocks are used all the way around. One set lasts all season. Incredible.

The rear axle itself is remarkably standard. Although the rules allow the use of full floating axle shafts (hubs fully supported by bearings at the end of the housing) the team no longer uses them. Floating axles are primarily a safety feature: If an axle breaks, the wheel doesn't come off. Penske axles don't break, the full floaters are heavier, and one such floater caused a leak that put one car out of a race, so they have gone back to the standard Chevy hardware. Hubs and axle shafts are crack-checked between every race anyway, and replaced as required. The normal Chevy positraction is fitted, with the clutches shimmed to tighten the differential action. This has a tendency to make it into a solid axle de-

vice, but should it slip excessively, the unit would quickly burn itself up at racing speeds. Differential coolers are allowed, but so far none have been required.

Brakes are the optional four-wheel disc systems, which utilize the Corvette assemblies. The usual power assist is fitted, along with a HD Corvette adjustable proportioning valve. Made by Kelsey-Hayes, it allows front/rear braking to be tailored to specific race courses. This is deliberately placed in an inconvenient place on the vacuum chamber so it will not be adjusted indiscriminately. Placing it within reach of the driver often allows him to proportion himself right out of braking effort. His job is to drive, not tinker.

The rules allow 9-in. wide wheels of any design and Penske uses the English Minilites. Yeah, yeah, we've heard all the stories about inferior British metallurgy, but these permanent mold magnesium designs have proven to be strong, light and cheaper than American designs. Though last year's wheels didn't break in combat, Donohue knew they were marginal. During the winter, he and Cantwell set up a test to find the strongest wheels available. The Minilites won. Now, we know that Penske's small shop doesn't have a fatigue test machine. Whose do you suppose they used? Their test was right, too. The only major team that didn't use Minilites was the Bud Moore Mustang crew. They lost two races and almost one driver because of wheel failure.

Because of the relatively narrow rim width, the latest wide tires have resulted in a decidedly pinched look. Firestone has brought its LXX design into the fray quite effectively for this reason, but so far Goodyear has not countered with anything exotic. (Penske is also a Goodyear distributor.)

Managing to fit the big gumballs under the fenders takes some real ingenuity. Scca says you can make all the room you need, so long as you don't take off any metal or add any, or recontour the fender. Thanks, guys. The internal wheel housings are expanded inward, some of the splash baffles moved out of the way, and then the outside fender is bowed out. This gets very tricky. It requires heating and beating it out, stretching the metal. The result is an extremely stock looking side slash that is noticeable only when parked next to a stock Camaro. Unfortunately these protrusions are the ones that catch the brunt of all the body contact that goes on in this kind of racing. They need rebending nearly every race, and tend to wear a little thin near the end of the season. In fact the entire fender gets ripplely. Acid dipping is illegal. Nobody said you couldn't grind off the wrinkles when repairing a dent. ("Daddy, what's a lightweight Camaro?" "Shut up and sand.")

One of the more interesting protests of the Trans-Am season involved the vinyl top used by the Camaros early in the year. It drove the competition, scca, and the visiting press absolutely batty, because nobody knew why it was there. Some claimed it was hiding an aluminum or fiberglass roof,

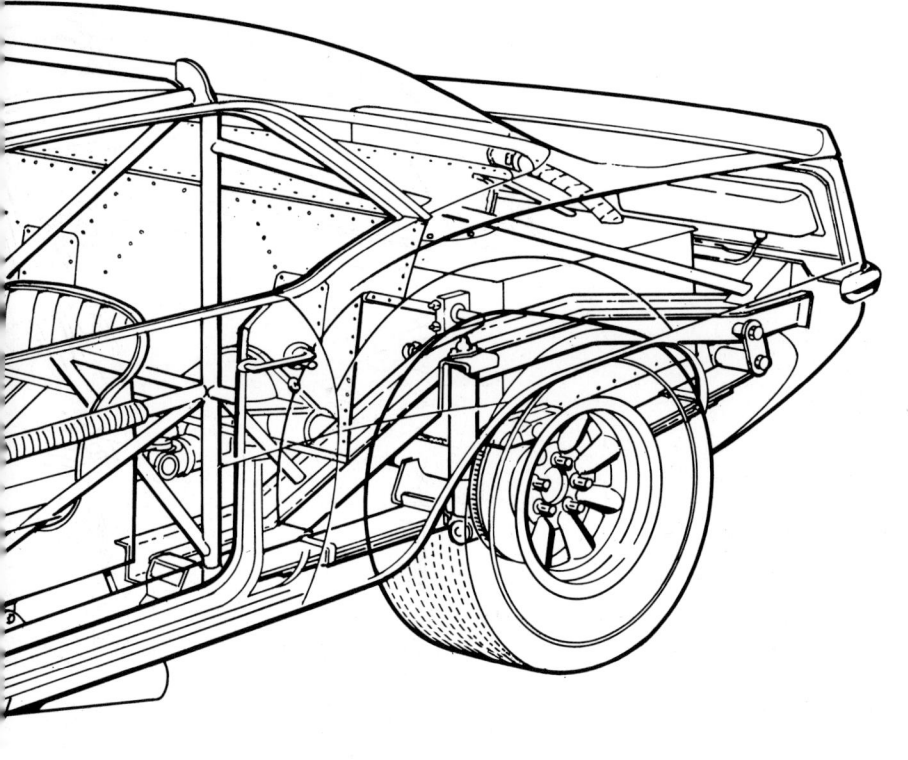

CUTAWAY BY DAVID KIMBLE

DIMENSIONS

Wheelbase, in. .108
Track f/r, in. .61/60
Ground clearance, in. 4
Curb wt (w/o fuel), lb. 2980
Distribution f/r 52/48

CHASSIS & SUSPENSION

Frame type: Unitized with separate front subframe.
Front suspension: Independent by S.L.A., coil springs, Koni adjustable racing shock absorbers, spec. 0.88 in., anti-roll bar. (Modified with longer spindle, eccentric plastic bushings.)
Rear suspension: Hotchkiss-type live axle, multi-leaf springs, Koni adjustable racing shocks, 0.50 in. anti-roll bar, Panhard track bar.
Steering: Recirculating ball, o/a ratio 17:1.
Brakes: Corvette 11.75 in. dia. disc, front & rear; vacuum assist; adjustable proportioning valve, Ferodo DS11 pads.
Wheels: 15x9 Minilite magnesium wheels.
Tires: Goodyear Blue Streak Racing, 5.00/11.30—15 front, (9-in. tread); 6.00/12.30—15 rear, (11-in. tread).

ENGINE

Type. .ohv V-8 by Traco
Bore x stroke, in. . . . . .4.02 x 3.00
Displacement, cu. in. 304.6
Compression ratio. 12:1
Fuel required. . . . . . .Sunoco 260 Super Premium
(103 octane)
Bhp @ rpm. . . . . .440 @ 7200
Carburetion: 2 x 4 Holley 600 cfm, cross-ram manifold.
Valve train: Mechanical lifters, H.D. pushrods, Microsealed H.D. rocker arms, Traco valves & springs.
Camshaft: Chevrolet high perf. (P.N. 3927140).
timing, deg. int./exh. 53-100/101-65
Duration: int./exh. 333/346
Lift, in. int./exh. 473/.492
Exhaust system: Steel tubing headers 1.75 dia., 36 in. tuned length; 3.5-in. dia. collector pipe, 44 in. tuned length.

DRIVE TRAIN

Clutch: Schiefer competition single dry disc.
Transmission: H.D. Close ratio 4-speed "rock crusher" (M22).
Gear Ratios: 1st. 2.20
2nd. 1.64
3rd. 1.27
4th. 1.00
Final drive: Hypoid w/Positrac limited slip. 12 axle ratios available, from 3.07:1 to 5.13:1.

PROFESSIONALS

although the tech inspector's magnets stuck to it. Others claimed it was some kind of super-slippery material that cut the drag.

Penske's story is that he ran them simply to set his cars apart from the crowd. Several other cars running the Trans-Am are also blue, including the Shelby Mustangs, and he simply wanted to look different for quick identification. Also the vinyl is tough, and would withstand several races before refinishing would be required, whereas paint must be touched up after every race, expensive even to Penske.

Curiously the most logical reason for using such a top was ignored. The forward part of a car roof is a negative pressure area. Venting the higher pressure cockpit air into that area would noticeably reduce lift. With a vinyl skin covering a series of pin holes, this could have been accomplished without the inspectors noticing. Nobody ever figured exactly *why* they ran the vinyl, though and finally got them off on a technicality. (The roofs didn't appear on the homologation papers.) The rules didn't say Penske couldn't use them, so his record was safe.

All engine work is left to Traco Engineering. Penske supplies the parts from his own dealership, Traco blueprints each engine, dyno tests it and keeps it tuned. Traco operates as an independent supplier, and is not on a retainer to either Penske or Chevrolet. Jim Travers does go to most of the races with Penske—as a paid pit crew member—and manages all the care and feeding of his babies.

Power figures are pretty subjective. By Traco's dyno, the engines average around 440 bhp at 7200 rpm. This is less than the other teams claim, but seems to get the job done. The Chevys are down on power to the Boss 302 Fords, but they win races anyway.

The engines use a surprising number of stock parts, including pistons, rods, cranks, sump, cam and pushrods. About the only thing made by Traco is the valve system, which offers a small improvement over the already good Chevy set-up. In anticipation of the dreaded rocker ball galling problem, Chevrolet homologated needle bearing rockers. However, Traco found it was just as effective and much cheaper to continue using the '68 system of Micro–sealing the standard rockers. Early engine failures were attributed to connecting rod weakness not encountered until the engine speed was increased to get more competitive

power. Subsequently the design was improved, shot peening added and the problem seems to be solved, *provided* they keep the revs below 7500, and don't try to get more than one race out of an engine.

They have found that it is false economy to try to get more than one race (plus practice) out of an engine between overhauls. (A full blow-up is costly, since it requires all new parts.) Overhauls consist of new rods, rings and a valve job. Everything is crack tested, including the block. The bearings are so good they are often used over again. Normally six engines are kept in the works: Two in the cars, two in the truck as spares, and two in overhaul.

A Schieffer flywheel and clutch along with a stock Chevy disc is used inside a safety bell housing. The M22 "rock crusher" transmission has the usual close ratio gears. A different set of gear ratios was homologated, but never produced, although they could have been effectively used on several courses.

In a form of racing where bleary-eyed mechanics always seem to be changing engines or gearboxes or rear axles right up to the starting grid, the Penske crew displays a remarkable picture of preparedness. At every race. The cars are ready when they arrive at the course. They have fresh paint, they are clean, they have the right jetting and gear ratios, and the mechanics are rested (well, relatively). There

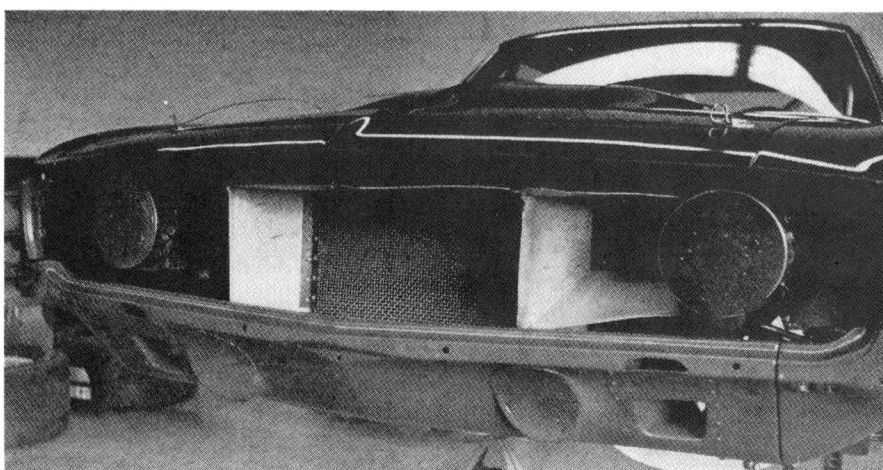

PREPARATION KNOWS no bounds. Grille removed reveals ducting and protective screen in front of Corvette aluminum radiator and Harrison oil cooler (**PN** 3157804 from your local Chevrolet dealer). Cars undergo complete overhaul between races, including repainting of even the undercarriage.

FENDER BULGING for tire clearance is kosher if contour is not changed, no metal is added or removed. Penske's cars stayed within the rules, retained an almost stock shape and still had room to spare. Internal wheel house is also expanded inward, splash guards removed or bent out of the way. Camaros are most stock appearing cars in Trans-Am.

is very little last ditch work done at the race course. Quiet, methodical preparation is carried out before the cars reach the track, so the crew can concentrate on going fast, not rebuilding cars.

Between each race, the cars are torn down to the "tub"—the unit-body and subframe. This assembly is steam cleaned, and repainted, inside and out, *including* underneath. The engine is sent to Traco, the suspension pieces are magnafluxed and the rear axle is exchanged for a fresh unit containing the proper ratio for the next race. Donohue and Cantwell determine what springs will probably be needed, based on previous experience or educated guess. The suspension goes back on using all new spindles, hubs, bearings and steering knuckles. Brake discs are replaced, as are the pads. All fluid systems are flushed, including the oil cooler and lines, radiator and hoses, and the fuel tank and pipes.

The transmission is replaced every two or three races as a precautionary move. A new engine goes in, the fluid systems are primed and the engine is ready to fire. The body work goes back on and any paint touch-up work is finished, including pinstriping. ("I want my cars sharp:" Penske.) The chassis is given a complete alignment, with anticipated changes incorporated. Donohue and Bucknum have a completely new car for every race, and they look it.

The race winning tactic this year was their pit stops, every bit as good as those of the legendary Woods Brothers. Each man has a specified job. If that certain job is not to be performed on a certain pit stop, then he stands back out of the way.

Sunoco designed a secret fueling rig, complete with cooling jacket containing dry ice and acetone. A huge hose and nozzle flow fuel at five gallons a second, about as fast as the car tank can take it. The spectacular fuel splashes common early in season have been all but eliminated by a better cap and vent system in the car. Seems Penske got a pair of $75 alligator shoes ruined by splashing fuel, hence the improved design. Tire changes are not normally anticipated in most of the races, but when and if, they are ready for those, too. Knock-off wheels are not allowed, so air wrenches and a five-bolt lug nut system are used. Getting them off is nothing. Getting a wheel on and off is helped by a special teflon-coated hub bushing. But fitting each individual nut was time consuming. Enter an ingenious device called the lug nut ejector. Five lug nuts are held in a cluster with the same spacing as the wheel. These are slipped over the studs, the ejector plate shoves them home, off the holder ready to be spun tight by an air wrench. Stripping of threads is kept to a minimum by coating everything liberally with spray Molybdenum.

Low on oil or water? One is hardly able to pour oil into a running engine or pull off a radiator cap on a pit stop. Hence air hose connections are connected to the oil and water systems. A pressurized cannister of oil or water is plugged in and the pressure in the cannister injects fluid into the hot and running engine.

What about 1970? For starters there will be an all new '70-½ chassis, with improved suspension design and fastback body style. No independent suspension, however. When you get right down to it, a well executed live axle on a big car is not all that much of a handicap, especially on the relatively smooth tracks. Semi-final 1970 rules subject to change at any moment have it that all manufacturers may destroke to arrive at the 305-cid limit, and that only one four-barrel may be used, *à la* NASCAR. The '70-½ Z/28 will therefore be a single carbed 350 (LT-1, of Corvette fame) with 370 street horsepower. Fitting last year's crank will bring it back down to 302. Will they need more horsepower? Doubt it. With only one carb, the air gulping ports of the Boss 302 will no longer give it the distinct advantage it had in '69, and the good torquing (relatively) Chevy should be back in the ball game again.

But instead of Sunoco Blue, the cars will be Chaparral White. A Texas Chevy story, J. Hall prop., will be campaigning them. With Turbo-Hydros, maybe?

But Penske, Donohue, Cantwell, all the Penske crew, and the sponsors will be wearing new uniforms. For that story, turn the page . . . ■

TRACO ENGINEERING has full charge of the engines. Penske owns six; two in the cars, two spares, and two at the shop being overhauled. One race only, between overhauls. Jim Travers follows the races, keeps the engines putting out the 440 bhp. Stock components are amazingly strong and well designed. Power is picked up in blueprinting.

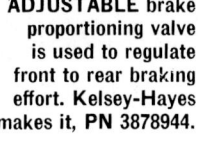

ADJUSTABLE brake proportioning valve is used to regulate front to rear braking effort. Kelsey-Hayes makes it, PN 3878944.

HUGE fuel tank filler neck has spring loaded trap door, allows flow of five gallons per second. Parallel operating Carter impeller fuel pumps deliver fuel to engine.

ROAD TEST
1970 CHEVROLET CAMARO

SCALE: 10" DIVISIONS

PRICE

List price, FOB Detroit..... $2839
List price, west coast...... $3002
Price as tested, west coast.. $4673
Price as tested SS package
(power brakes, 300-bhp engine,
F70 tires, 14x7 wheels—$290),
handling option ($31), RS package ($168), Turbo Hydra-Matic
($222), power steering ($105),
A/C ($380), extra instruments
($84), etc

MANUFACTURER

Chevrolet Division of General Motors Corp.
GM Bldg., Detroit, Mich. 48202.

ENGINE

Type...................... V-8, ohv
Bore x stroke, mm.... 101.6 x 88.4
    Equivalent in....... 4.00 x 3.48
Displacement, cc/cu in.. 5735/350
Compression ratio........ 10.25:1
Bhp @ rpm........... 300 @ 4800
    Equivalent mph............ 110
Torque @ rpm........ 380 @ 3200
    Equivalent mph............. 75
Carburetion.... one Rochester (4V)
Type fuel required...... premium
Emission control.... engine mods

DRIVE TRAIN

Transmission... 3-speed automatic
(torque converter with planetary
gearbox)
Gear ratios: 3rd (1.00)..... 3.07:1
    2nd (1.52)................ 4.67:1
    1st (2.52)................ 7.74:1
    1st (2.52 x 2.1)......... 16.28:1
Final drive ratio........... 3.07:1

CHASSIS & BODY

Layout.... front engine, rear drive
Body/frame: unit steel with front
subframe
Brake type: 11.0-in. vented disc
front, 9.5 x 2.0-in. drum rear;
vacuum assisted
    Swept area, sq in.......... 332
Wheels......... steel disc, 14 x 7
Tires. Firestone W.O. belted F70-14
Steering type: recirculating ball,
power assisted
    Overall ratio....... 15.5-11.8:1
    Turns, lock-to-lock.......... 2.3
    Turning circle, ft........... 41.1
Front suspension: unequal-length
A-arms, coil springs, tube shocks,
anti-roll bar
Rear suspension: live axle on leaf
springs, tube shocks, anti-roll
bar

ACCOMMODATION

Seating capacity, persons ... 2+2
Seat width,
    front/rear..... 2 x 22.5/2 x 20.0
Head room, front/rear... 37.0/35.0
Seat back adjustment, degrees...0
Driver comfort rating (scale of 100):
    Driver 69 in. tall............ 90
    Driver 72 in. tall............ 70
    Driver 75 in. tall............ 70

INSTRUMENTATION

Instruments: 150-mph speedo,
7000-rpm tach, 99,999.9 odo,
water temp, ammeter, fuel level,
clock
Warning lights: oil pressure, brake
system, high beam, directionals

MAINTENANCE

Service intervals, mi:
    Oil change................ 6000
    Filter change........... 12,000
    Chassis lube............. 6000
    Minor tuneup............ 6000
    Major tuneup........... 12,000
    Warranty, mo/mi 12/12,000 (car)
    60/50,000 (powertrain)

GENERAL

Curb weight, lb........... 3670
Test weight............... 4045
Weight distribution (with
    driver), front/rear, %.... 57/43
Wheelbase, in............. 108.0
Track, front/rear...... 61.3/60.0
Overall length............ 188.0
    Width.................. 74.4
    Height................. 50.5
Ground clearance............ 4.5
Overhang, front/rear.... 38.1/41.9
Usable trunk space, cu ft..... 6.5
Fuel tank capacity, U.S. gal... 18.0

CALCULATED DATA

Lb/bhp (test weight)........ 13.5
Mph/1000 rpm (3rd gear).... 23.4
Engine revs/mi (60 mph).... 2560
Engine speed @ 70 mph.... 2970
Piston travel, ft/mi........ 1485
Cu ft/ton mi.............. 128.5
R&T wear index............. 38
R&T steering index......... 0.94
Brake swept area sq in/ton.... 164

ROAD TEST RESULTS

ACCELERATION

Time to distance, sec:
    0–100 ft................. 3.8
    0–250 ft................. 6.2
    0–500 ft................. 9.3
    0–750 ft................ 11.9
    0–1000 ft............... 14.1
    0–1320 ft (¼ mi)........ 16.6
Speed at end of ¼ mi, mph.... 86
Time to speed, sec:
    0–30 mph................ 3.5
    0–40 mph................ 5.0
    0–50 mph................ 6.7
    0–60 mph................ 8.8
    0–70 mph............... 11.2
    0–80 mph............... 14.4
    0–100 mph.............. 24.4
Passing exposure time, sec:
    To pass car going 50 mph.... 6.0

FUEL CONSUMPTION

Normal driving, mpg........ 14.4
Cruising range, mi.......... 251

SPEEDS IN GEARS

3rd gear (5000 rpm)......... 115
2nd (5000)................. 74
1st (5000)................. 43

BRAKES

Panic stop from 80 mph:
    Max. deceleration rate, % g.. 84
    Stopping distance, ft...... 281
    Control........... excellent
Fade test: percent increase in pedal
    effort to maintain 50%-g deceleration rate in 6 stops from 60
    mph...................... 270
Parking: Hold 30% grade?..... yes
Overall brake rating........ fair

SPEEDOMETER ERROR

30 mph indicated is actually.. 28.5
40 mph................... 38.6
60 mph................... 58.6
70 mph................... 68.5
80 mph................... 78.5
100 mph.................. 98.7
Odometer, 10.0 mi.......... 9.8

ACCELERATION & COASTING

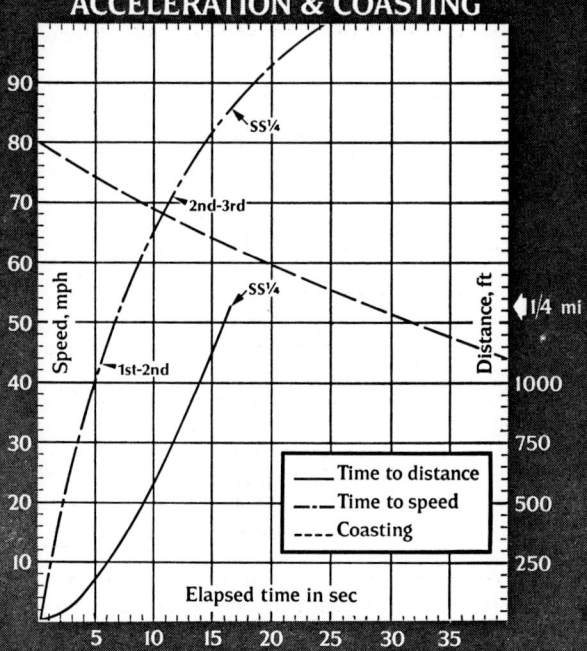

Speed, mph | Distance, ft
Elapsed time in sec

SS¼
2nd-3rd
1st-2nd
SS¼
¼ mi

— Time to distance
-·- Time to speed
--- Coasting

94

1970 CHEVROLET CAMARO

*A tremendous improvement—puts the Ponycar in a new class
... but the brakes aren't up to the weight or performance*

THE NEW CAMARO is, except for its engine and drivetrain, an entirely new car and represents what we think is the first serious effort since the 1963 Corvette to create a real American GT. Substantial and meaningful changes have been wrought on the Camaro's chassis—greatly improved suspension, front disc brakes as standard equipment—and though the stylists have continued to reign supreme in the body layout, the new model is esthetically successful and clearly more comfortable than the old.

We got our test car about a week before public introduction and greatly enjoyed the reactions of people on the streets to it. Some practically crashed into trees gawking at its European snout and graceful lines, but we got the feeling several times that drivers of older Camaros were purposely ignoring it. Did they feel abandoned, or did they simply not realize it was a Camaro?

The choice of options on our test car was quite important, as it always is on American cars. We shied away from the striped, spoilered, race-geared Z-28 after trying one: a Z-28 may be a great thing to be seen in at the local discotheque

and it's nice if you want to identify with Trans-Am racing, but it's not the way to go if you want to travel long distances fast and in comfort. And after all, that's what *Gran Turismo* is all about. Our car, then, was a Rally Sport (not an important fact because the main item in the RS package is the plastic snout and split front bumpers) with the 350-cu-in. 300-bhp engine, chosen because it's the most powerful engine available without getting noisy mechanical valve lifters and a rough idle or the 180 lb extra weight of the big-block 402 engine. This was tied to the excellent Turbo Hydra-Matic transmission, because we object to the stiff Hurst shift linkage used with the 4-speed box, and a 3.07:1 final drive for relaxed high-speed cruising. In the chassis department the test car had the SS (Super Sports) and handling packages: a larger-than-standard (1.0-in.) front anti-roll bar, the new rear anti-roll bar, 7-in. rims (vs the standard 6-in.) and F70-14 belted tires. Finally, to complete the GT theme we got the optional instrumentation, which includes a tachometer and extra gauges, and air conditioning.

Behind the wheel we were impressed by the much improved driving position—the seat has more travel so the

CHEVROLET CAMARO

wheel can be farther away than before—and the nice view out over the hood. Vision to the front is good, to the rear not so good but the blind rear quarters aren't as bad as we expected because they are well forward and can't completely hide a car behind and to the right. Two outside rearview mirrors also help. The optional instruments are laid out in classic GT style: large (but not too large) speedometer and tachometer, flanked by three minor gauges and a clock angled toward the driver. There's no oil-pressure gauge and we're just old-fashioned enough to want one; we also missed a trip odometer. The controls are not outstanding; there are no steering-column stalks, to facilitate things like operating windshield wipers and washers without fumbling or looking, as found on many European GTs. Two things are noteworthy: an optional remote-control outside rearview mirror for the driver (why don't we get these on Europeans?) and a tilting steering wheel, also optional.

The heating, ventilation and air conditioning controls rate a separate description. Their master panel, over to the left of the dash and low, is nicely lighted at night but obscured from the view of most drivers by the steering-wheel spoke; there's a 4-speed blower which, with A/C, can't be shut off as Chevrolet air conditioning partially disables the normal ventilation system. The A/C itself has settings for fresh, recirculating and "bi-level" air, the latter of which gives warmer air to the feet than to the multitudinous dash vents for those days when it's cool outside but sunny. Weather during our test wasn't adverse enough to evaluate any of these but they certainly look impressive!

The interior of our test car was luxurious and generally tasteful. The front seats, however, are barely acceptable; they don't provide lateral support in keeping with the Camaro's cornering ability even if the driver has his lap and shoulder belts on. (These still have two separate buckles, the least satisfactory arrangement available.) And the backrests are not adjustable, though we've heard that such will be available later. Rear seating is strictly of the +2 variety despite the car's large size—the cushions are very low and the leg-room minimal, which means knees-in-the-air. Access to the rear seat is good, thanks to the wide, wide doors which with their built-in impact bars must weigh about a ton apiece. And as for trunk space—well, 6.5 cu ft is on a par with an Opel GT or Porsche 911 and we had to allow for a little box in every nook and cranny to come up with even 6.5. A Space-Saver spare tire is optional and recommended.

That 350-inch engine and automatic transmission are the kind of things the Americans do best—a marvellous combination of mechanical quietness, nice throaty exhaust beat and lack of fuss. Sure, the hot Z-28 engine puts out more power and revs higher, but high specific output and revs aren't an end in themselves and, since this level of output

CHEVROLET CAMARO

gives more than adequate performance for all road conditions to be encountered in America, we favor this "mild" 300-bhp engine. We also disdain the heavier 402 or 454 for general use, as the Camaro is already so nose-heavy as to give marginal traction in rain, snow or ice.

The Turbo Hydra-Matic shifts smoothly and decisively—a rare combination of qualities—and once one is accustomed to the notches of the console shifter the intermediate gear can be held or selected for engine braking without undue concentration. Even with the automatic box fuel economy, at 14½ mpg, isn't oppressive but the 18-gallon tank makes for a limited range between refills. We also wonder how long "the masses" can go on in this country using petroleum at this rate before petroleum reserves are seriously depleted or all our beaches are black . . .

In our road test of the Camaro Z-28 two years ago we classified its ride as "clumpity clump." One of the objectives of the Chevrolet engineers in developing the new Camaro was to avoid this in cars equipped with the "handling package," which that Z-28 and the present test car both had. Softer rear springs, with a bigger front anti-roll bar and a new rear anti-roll bar to restore roll resistance, do make quite an improvement in the "handling" Camaro's ride. But the ride is still on the stiff and jiggly side. Increased rear suspension travel has lessened the tendency, previously serious, for the rear end to bottom on large bumps and dips and the new car, still not outstanding in this respect, is nevertheless better than any other Ponycar we've driven. Over patched-up pavement the wide bias-belt tires cause a lot of wandering; radials would eliminate this.

The new power steering is eminently successful, though. It is very quick—15.5:1 at the center, decreasing to 11.8:1 at the locks and requiring only 2.3 turns lock-to-lock—and it gives a degree of natural steering feel heretofore unknown in domestic power steering. In fact it's quite close to being as good as our ideal (the Mercedes-Benz device) and the quick, variable ratio makes the Camaro uncommonly maneuverable for such a long-hooded beast.

To assess the effects of the suspension changes we evaluated the Camaro's handling on both smooth and rough pavement. On the former the difference is striking: the rear anti-roll bar provides more roll resistance at the rear than did the former stiff springs, meaning that there is less understeer than before. This shows up as soon as you head the Camaro into a bend; now it's possible to get it into a neutral cornering attitude by a mere tweak of the steering wheel and then keep it there with the throttle. Before, the only way to overcome the built-in understeer of so nose-heavy a car was to break the tail loose with a lot of throttle opening, and this can be a tricky proposition. On rough roads the softer spring rates keep the rear wheels a bit more dependably planted on the ground, but not much—a live rear axle with lots of roll resistance and a big engine feeding it with torque just isn't very determined to stay in contact with a rough road. Speaking of rough roads, the body is rigid but detail rattles and squeaks are there, particularly around the doors which seem to be "working" constantly against the door seals (a new type which prevents any trace of wind noise) even on relatively smooth roads.

As mentioned earlier, disc front brakes are now standard, a nice step forward; but their swept area has not been increased to match the Camaro's weight increase with the result that fade resistance, at least on the car we tested, is pretty terrible. In fact, in our six-stop fade test the pedal effort required to maintain ½ g deceleration increased from a modest 27 lb to a hefty 100 lb—actually the brakes wouldn't quite hold ½ g at the end of the sixth stop. By contrast the Camaro we tested in 1967, weighing 3240 lb and using the same brake system, got through the six stops with only 34% increase in pedal effort and our Z-28 of 1968 at 3355 lb (presumably with harder pads and linings) faded not at all. In the panic stop from 80 mph our 1970 car did very well, stopping in a commendable 281 ft with excellent directional control. And in everyday use the feel of the power-assisted brakes is quite good. In other words, these brakes are adequate for everyday driving and they will stop you nicely in an emergency from 80 mph, but you wouldn't want to use them repeatedly, say, in vigorous mountain driving. Four-wheel discs, which the car really needs, are not available, more's the pity.

If we owned a Camaro like this we'd replace the front pads and rear linings with something harder. Other than the brakes, our only serious objections to the new Camaro arise from something that is common to all American cars: it's too big for the cargo of people and luggage it carries, and it doesn't offer the latest and greatest combination of ride and handling. But overall it's a pleasant, responsive, solid car—very nice to drive in the day-in-day-out routine and an exceedingly good long-distance touring car. In fact, we'll have to say it's the best American car we've ever driven, and more importantly it's one of the most satisfying cars for all-around use we've ever driven. ⊙

ZAPF

THE FEEL AT THE WHEEL

BY DAVE EPPERSON

The idea of driving another squishy piece of Deetroit pigiron, then trying to discover what, for publication, *isn't* wrong with the big turkey usually induces a twinge of nausea in the pit of of a magazine test driver's stomach. At the typewriter, later, the road tester stares blankly at the paper, wondering what he's going to write about, wondering if the Detroit manufacturers ever will listen to the enthusiasts' plea — better suspension, better tires, better transmissions, quicker steering, and to hell with the gutless wonder and its "boulevard ride," so called.

"It's just another Chevy Camaro — a little late for '70 maybe — SS, Z28 and all that flakey hoak. So what? Gimmie the keys."

The console comes up tight to the right thigh and the door snugs shut, securing the driver as much as belt and shoulder strap. The polished ball of the 4-speed gearbox shift lever eases into the cup of the right hand. Clutch, brake, throttle — all pedals are there, in reasonable order. The ignition key slides into the steering column mounted lock. A couple of whirring revolutions off the new-car-fresh battery and the Z28 special 350-cu. in./360 bhp engine bubbles to life.

"Sounds like any 'Vette. So who's impressed."

Exhaust purrs like a leopard. The big four-barrel Holley's induction hiss is thinly disguised behind minimal air cleaner restriction. A blip of the throttle wings the tach's red needle to five grand. The engine makes no complaint.

"They've made it sound like it means business, but man, how many times you heard, 'If it won't go, make it loud and chrome it.'?"

A firm arm-wrestle snicks the gear lever past the gate into reverse. Clutch action smooths the Camaro out of its parallel parking space. First gear is a short-throw flick. Even in the parking lot, the Z28 begins to exhibit a sort of flexible, trundling, flat-as-a-board feel.

"Oh, yeah? Well, stiff suspension does not a sports car make."

Third gear is fine for trickling through city traffic. The 350's torque curve and 3.73:1 rear axle lets the Z28 slide along anywhere from 20 to 50 mph without an up or downshift. The freeway on-ramp invites the test driver to now downshift into second gear, punch the throttle, then change up to third, then fling it into fourth to achieve escape velocity.

"Whoops! That's 95 mph, friend. A guy could get arrested doing his thing on a public turnpike."

The grin comes automatically as the Z28 eases to legal speed, transmitting essence of road feel through taut suspension and positive steering. The Camaro tracks precisely along the sight-line chosen by the driver — an exact 6 in. from the road-edge seam, or 3 in., whatever the driver chooses.

"Lots of cars do reasonably well on freeways, but go to hell in a hand-basket at the first sign of a fun piece of roadway. In the past, Camaros, even the original Z28s weren't much different than the rest. Disappointment time."

The driver peels off the freeway

ED BY A Z28 CAMARO

CORVETTE USED TO BE CHEVY'S SPORTS CAR BUT NOW IT'S CAMARO TO CARRY THE COLORS

CAMARO

and heads for his isolated private stock of crooked road, a can-of-worms section that has no access roads, a safe snakey hank of ups and downs, hairpins and off-camber bends, with unobstructed vision of two miles or more in some places, a secluded skein of asphalt made to put cars to the test.

Accelerating into the first bend, a glance at the speedometer shows, surprisingly, a reading 20-odd mph above the "feel" speed of the car. And the tach, too, is reading higher than the sound of engine rpm. The first big sweeper swings easily under the wide stance of the Z28. Brakes; downshift; throttle! Booming through a nasty little hooker at the bottom of the hill, the Camaro feels as though it has claws for digging into pavement, rather than Goodyears.

Another run and another and another along the remote roadway increases the degree of man-car/car-man oneness, the beautiful familiarity between human and machine. Suddenly, sunset says the afternoon and a tank of fuel have vanished.

"Okay! Okay! Okay! I take back everything I ever said. This cotton-picking Z28 Camaro is the best out-of-Deetroit, factory car I ever drove."

To find out why a hardened test driver would make such a broad, if not rash, statement, it's necessary to go back a few years — when Mustang was about to become 1,000,000-unit seller, and GM's Chevrolet people had yet to produce a Ponycar. Everybody who knew the automobile business figured the GM Ponycar, when it emerged, would be simply a Chevy II core wrapped in computer-designed sheet-

metal. The experts weren't far from wrong. Sub-frame and unit body, plus leaf springs at the rear, added up to Chevy II, inside, sort of.

The sheetmetal outside was smooth, undistinguished, unobtrusive and unimaginative, sort of. The original Camaro looked like what it was — the Chevrolet Mustang.

"I'd like to shake the hand of the guys who worked up this car."

Fortunately for Chevylovers everywhere there were (still are) two men in Michigan, Zora Arkus-Duntov and Walt MacKenzie, who realized that some Chevy people want more than a kissyersister kind of car. Product of Duntov engineering and MacKenzie promotion, one of the first Z28 Camaros showed up at GM torture track near Mesa, Ariz., early in 1966. This car had a just-over-300-cu.-in. engine, fiddled suspension and fair brakes. It was, for its time, an exciting car.

That initial Z28 showed more potential than out-and-out performance. That Z28 was designed primarily for the good old American sport of production sedan racing (imported from good old Great Britain). Furthermore, in years that followed, some Camaros were campaigned with a good measure of success in local, regional and international sedan class events, including the famed and furious Trans-Am series. On the track, people with and without backdoor aid from Chevrolet were doing things to Camaro suspension, ridding it of roll, dive, sway and other mannerisms not compatible with serious road racing.

Meanwhile, back in Deetroit-city, Duntov continued to tweak up Corvette engine — 350s and 427s and such like. These grew in horsepower, torque, reliability and free-breathing, high-

winding competition capability. And, chassis engineers and stylists, operating on the realization that it wasn't quite right on the first go-around, set about to re-do the Camaro.

"Y'know they've really worked this baby over!"

Eventually, the whole thing, suspension technique gained on the Trans-Am circuit, plus Duntov's super-strong 350 'Vette engine, plus a healthy dose of chassis engineering, plus some careful refinement in exterior styling, all are distilled into the 1970 Z28 Camaro.

The new Camaro's sub-frame uses larger box sections in side rails and a single, large crossmember, rather than two smaller crosspieces. Front suspension attached to the frame members shows increased span in both upper and lower A-arm bushings. Loads are reduced by increasing ball joint span to 1.4 in. A large diameter anti-roll bar is installed.

Rear suspension still uses live axle and multi-leaf semi-elliptic springs. Spring rates have been reduced for a more flexible ride, yet wheel control is improved through higher rates in shock absorber damping. Optional high performance suspension systems include higher rate shock absorbers and an anti-roll bar system for added stability. Each optional engine installed at the factory is given its own "tuned" suspension system.

In the test Z28, the tuning was as flawless as a concert by The Mamas And The Papas. The engineering that produced wider track, modified suspension, a variable steering ratio that seemed superquick, and better weight distribution has resulted in vastly improved handling.

"It's the closest thing to a competi-

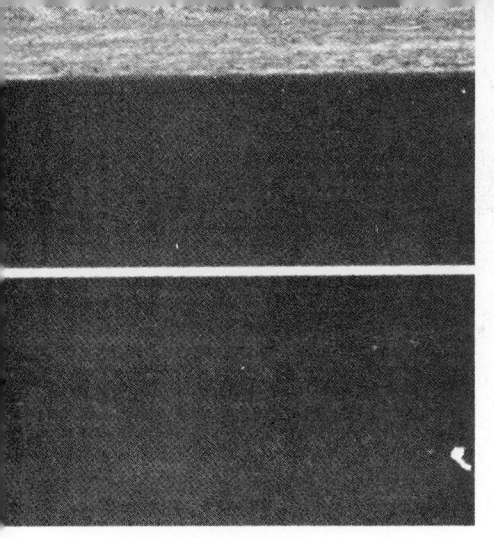

tion I ever drove on the street."

If the Z28 Camaro is the ultimate among U.S. built high performance road cars, it also performs well in the quick and fast categories of quarter-mile acceleration. Off the line, the big →

CAMARO

clutch can be feathered home to minimize wheelspin, then banged in-out for the one-two and three-four power-shifts for seemingly uninterrupted, pure acceleration to and through the lights. ET of 14.5 and top speed of 98.2 tell the story.

The 250-cu. in. Six (rather than the 230) is Camaro standard. Optional powerplants are the 307-cu. in./200-bhp, the 350-cu. in./300- and 360-bhp, and (later this year) the 454-cu. in./450-bhp V-8s. The 396-cu.in./325-bhp V-8 has been discontinued, but a 396-cu. in./375-bhp powerplant remains available for the Camaro. The 302-cu. in. V-8 has been replaced by the 350/360 for the special Z28 package.

"This isn't a drag racer, but it'll do in a pinch!"

By this time, the test driver has taken to patting the Z28 Camaro on the flank, or running his hand along the easy curve of the car's fender line.

Aircraft inspired airscoop forward, broad competition stripes on hood and deck, mag-type wheels, tires that lay down an 8.5-in. wide pad, and uptilted spoiler give the Z28 an appearance that matches its performance. The viewer has the feeling that this is the way a Lamborghini would look if it were meant to run on U.S. interstate routes and speed limited two-laners.

"Yeah, this is the kind of car that ought to make Jaguar eat its heart out. The Camar'll do everything a Jag will do, and about 30 percent quicker, surer and faster."

As an attention getter, the Z28 has no peer. MOTORCADE's test car was on the street some weeks prior to official introduction. Parked, it invariably drew a knot of tire kickers, all approving. On the road, it drew obviously admiring glances wherever it went. One enthusiast, piloting a seen-better-days Caterpillar yellow MGA grinned and offered the thumbs-up sign. A blonde in an Opel station wagon beside the Camaro at a stop-light called out with a smile, "My, that's a nice looking car." One car nut who had placed an order for Hugger Orange Chevrolet El Camino took one look at the Camaro, then hustled off to his dealer to change that order.

"I'm going to buy me one of these at the end of the model year when they cut prices – and have some of the bugs worked out of assembly."

The test driver's threat to buy a car such as the one he's just been testing is almost unheard of in the magazine business. Bugs in new model runs, however, are not.

In the case of the Z28 Camaro supplied to MOTORCADE, the bugs were few. The throttle pedal first jammed at about half throttle, then fell off altogether, which wasn't entirely bad because it left the driver a smooth metal rod on which to press his foot, giving excellent power control. The gear shift knob spun loose and jingled with transmission vibration; a click-click squeak developed in the right windshield pillar; but, for the most part, this Camaro showed much better than average assembly.

"Hey, man, I really hated to see that one go back. How soon do you figure we'll be able to get away with testing another Z28 Camaro?"

TEST CAR:	CHEVROLET CAMARO Z28

PRICE

Basic list, FOB	$2,839
As tested	4,420

Options included: Rally sport package, Z28 performance options, four-speed transmission, limited slip, power brakes, power steering, console, tinted glass, special instruments, AM radio and rear speaker, tilting wheel, outside mirrors.

ENGINE

Type	ohv V-8
Bore x stroke, in.	4.00 x 3.48
Displacement, cu. in.	350
Compression ratio	11.0:1
Rated bhp @ rpm	360 @ 6000
Rated torque @ rpm	380 @ 4000
Induction system	1 x 4 Holley
Electrical system	12 V 61 amp hr.
Type fuel required	Premium

DRIVE TRAIN

Transmission type	Four-speed manual
Clutch diameter, in.	11.0
Gear ratios: 1st	2.52
2nd	1.88
3rd	1.46
Shift lever location	console
Differential type	hypoid, limited slip
Axle ratio	3.73:1

CHASSIS AND SUSPENSION

Frame type	unitized with stub frame
Brake type	11-in. disc fr./9.5-in. drum, power assisted
Brake swept area, sq. in.	332.4
Steering type	recirculating ball, power assisted
Turns lock to lock	2.3
Turning circle, ft.	38.9

Front suspension: Unequal length upper and lower A arms, coil springs, tubular shock absorbers, anti-roll bar

Rear suspension: Multi-leaf semi-elliptic leaf springs, tubular shock absorbers, anti-roll bar.

DIMENSIONS AND CAPACITIES

Wheelbase, in.	108.0
Track, F/R, in.	61.3/60.0
Overall length, in.	188.0
Overall height, in.	50.5
Overall width, in.	74.4
Curb weight, lb.	3550
Weight distribution, F/R	57/43
Ground clearance, in.	4.5
Front seat hip room, in.	2 x 22.8
Shoulder room, in.	56.7
Head room, in.	37.4
Pedal-seatback, max. in.	40.5
Rear seat hip room, in.	47.3
Shoulder room, in.	54.4
Leg room, in.	29.6
Head room. in.	27.4
No. of passengers	4
Luggage space, cu. ft.	7.3
Crankcase, qt.	4
Cooling system, qt.	16
Fuel capacity, gal.	18

WHEELS AND TIRES

Wheel size and type	15 x 7
Optional size	14 x 6, 14 x 7
Tire size and type	F60-15 Goodyear Polyglas
Normal inflation, F/R, psi	24/26
Maximum load per tire	1500 @ 32

FUEL CONSUMPTION

Test conditions, mpg	10.6
Normal conditions, mpg	11.5
Cruising range, mi.	190-210

PERFORMANCE

Standing ¼-mile, sec.	14.5
Speed at end, mph	98.2
Top speed, mph (est.)	125
Braking performance	Excellent
Directional control	Excellent
Rpm redline	6500

NEW AND IMPROVED

The Camaro SS approaches GT status

With few exceptions it is difficult to select an American car for evaluation that has pure enough blood lines to elicit response from owners, a companion feature to this road test. One immediately thinks of the Corvette as the only ''pur sang'' domestic, but there are others, currently the ''ponies,'' that have changed little in aspect since introduction. We chose the Camaro for assessment for several reasons, not the least of which was the mid-winter introduction of the all new body and major changes underneath. And we feel that the Camaro has changed little in spirit and mainly for the better since its debut in late 1966.

There is no doubt that Mustang, first on the market, was the harbinger of the specialty cars that now take their nickname from the originator and are called ''ponies.'' Mustang had a two-year edge on the Chevy,

and at introduction the Camaro suffered some by comparison, plus the origin of the name came in for a good bit of unflattering conjecture. The first Camaro was hailed by enthusiast magazines as a really new car; some thought perhaps it might evolve to a four-passenger Corvette. Writers remarked on the styling similarity to the Corvair, and the fact that the general parts bins of Chevelle and Chevy II had supplied most of the components. A variety of sixes and V-8 engines was available with the 350 cu. in., 295 hp V-8 at the top of the heap, although the big six and the 327 V-8 were more popular. Camaros were available with the two-speed automatic (Powerglide) or three- or four-speed manual transmissions. The long option list included trim packages and disc brakes for the front. Inside the dash layout and hardware were frankly borrowed from Corvair; out-

side the rakish ''Coke'' bottle styling bore the proper family resemblance, and the car was more of a small size sedan than the sports machine implied in the ads.

Introduced with Cougar and fighting the established Mustang and Barracuda, the Camaro was not a sales phenomenon. However, in just a few months the Z-28 was born and homologated for racing under Group 2, FIA. The Z-28 had a lightweight body, the destroked 327 sized at 302 cu. in., performance packages in the form of power disc brakes, heavy duty suspension, rear spoiler, and of course, the inevitable racing stripes. The performance image caught on, and Camaro climbed the sales ladder.

Taking a tip from the import invasion, the makers of ''pony'' cars contented themselves with nominal changes in styling from year to year. As the factories got deeper and deeper into the performance game in the late sixties, the option lists escalated like inflation. The builders of Camaro were right in the middle of the option antics, carefully tailoring special performance packages to be competitive everywhere from the drag strip to the road racing courses of the Trans-American Championship. The Z-28 acquired more and better spoilers, suspension options, four-wheel disc brakes, etc. The engine list rose to include Chevy's big 'uns for the drag racers. The Z-28, 302 inch engine produced for little else than to win Trans-Am races did just that, securing the Championship in 1968 in a walk-away, and winning out over Mustang in 1969 after a hard fought season. Other Camaros did well at the drag strip, and many other racing cars used the 302 Chevy with great success—the most notable being in the SCCA's open wheel championship for Formula A cars powered by American made, five liter V-8s. All but a very few power-plants in Formula A were and still are Chevy.

Camaro. The other major changes will be explored as we get further into the story. We did feel that the Z-28, while a popular model, was a pole or two away from the great volume-sales Camaro. So we decided to try a different model. Our first thought was toward the middle ground, 350 inch engined number, but then nearly every magazine had been that route. The one Camaro generally ignored by the road testers is the SS 396, reputed to be a bad actor because of the heavier engine in the nose. Our test SS 396 also was equipped with nearly every accessory known to man and the option writers which brought the total cost from the base of $2839 to a whopping $4977, plus tax and license. Now the copy writers have been heavy on the idea, stemming from the hugger promotion days, of the Camaro graduating from a sport sedan to a Grand Touring machine. The price tag alone qualifies the test car for the GT category, so we set about to find the other qualities so

necessary in a car for touring in the grand manner.

What's Under the Hood

The new Camaro has a longer hood than its ancestors; in fact, the cockpit is some three inches further toward the rear this year, and the length goes on the nose. It would appear that the long hood is needed though, for a variety of engine packages can be stuffed into the bay. Our test car was equipped with the largest engine available for the Camaro, the Turbo-Jet 396 which is really 402 cubic inches in displacement derived from a bore of 4.126 inches and a stroke of 3.76 inches. The 10.25 to 1 compression ratio is for high performance, and this Camaro most definitely drinks premium gas only through a single four-barrel carburetor. The 396 is rated at a conservative 350 hp; that figure is no doubt intended as a sop to the owner's insurance rates. The husky torque figure of 415 lb./

The 1970 Camaro has a new profile with a full slant from the roof to the tail and a pointy, but lower nose, and wider doors.

The rear of the Camaro is clean in styling with a short Kamm-type tail adorned with Chevy style round tail lights.

Still the march of progress goes on, and the minor face lifts on the Camaro gave way in 1970 to a really new looking car. Industry rumors claimed the new Camaro, introduced in February, was intended to be a 1971 model, but sales pressures and other factors brought the car to the consumer six months early. There have been plenty of changes! *Road Test* magazine evaluated the Z-28 in the June issue, so faithful readers know that the distinctive, race oriented 302 engine is no longer an option for the

ft. is more realistic and representative of the urge available from the engine.

The SS 396 power is coupled to a four-speed manual transmission with the gear lever comfortably mounted on the center console. The fairly wide ratios of the four-speed work well with the 3.31 rear axle standard with this engine. The gearbox itself seemed a bit notchy and stiff with a tendency to hang up when the driver went looking for first gear. After experiencing the butter-smooth Corvette gearbox, the stiff, almost new-British feel in the Hurst linkage was a surprise, and the stiffness did not lessen as the car built up break-in miles. While on the gearbox subject, let us digress a bit. Now we appreciate the manufacturer's desire to make the car theft proof, but we would prefer a simple interior release for the engine hood (to protect the expensive powerplant) to the infernal mechanism that locks the steering wheel and releases the key when reverse gear is engaged. Several times we were tempted to leave the Camaro four-speed parked with the key dangling hopelessly in the ignition after struggling manfully to push, cram, or jam the balky lever into reverse in order to get the ignition key released. On the same vein we dislike having to start a car with the clutch depressed, and that is the only way the car will fire with the safety gadgetry installed. It is particularly awkward on a hill where one must either heel and toe (not easy either) to keep from sliding backward, or trust the marginal parking brake to hold the car. We suppose that eventually even high performance GTs will be available only with automatic transmission thus eliminating these problems, but it is an annoyance to the enthusiastic driver who still stubbornly prefers to shift for himself.

The new additions to the Camaro

suspension have been explored before, and the handling package of the SS designation is part of the new GT image for 1970. On all models the front end has a wider tread by 1.68 inches over earlier models. Along with this the upper and lower control arm bushings have an increased span as does the ball joint. At the rear live axle, multiple leaf springs replace the original single leaf set-up, and new rear spring bushings, wider rear tread, etc., improve the ride. The handling package includes hefty anti-roll bars both front and rear, and these really make the difference for the nose heavy 396 whose engine weighs nearly 200 pounds more than its milder brethren.

Happily, Chevrolet has made the front disc brakes a standard item on all Camaros, and the units are burly 11 inch vented discs. At the rear, standard drums are fitted and the whole works has a pleasant and pro-

gressive feeling power boost. Camaro has also acquired the GM variable ratio power steering which provides much better road feel than previous power systems, and the recirculating ball type steering is more forward mounted under the redesigned nose. Not included in the SS group, but supplied on our test car was the very valuable Positraction option, and the limited slip diff adds more aura to the GT aspect of the Camaro. There were no radical suspension changes for the 1970 Camaro, but subtle improvements add to the ride quality, particularly to the SS models whose stiff springing and fat tires formerly yielded a rather harsh ride on smooth roads.

Styling and Appointments

The stylist is king in Detroit and the engineering often seems a secondary consideration when a car evolves through a major change. The Camaro, fortunately, comes off with its mechanical improvements unhampered by the sleek new body that is faintly reminiscent of a well known GT from Italy. The extremely long hood, low roof line, and faired in "coke" bottle flanks end abruptly in a Kamm type tail adorned with the familiar round Chevy tail lights. The styling is unquestionably good, and European influenced, and all but the real traditionalists will think it more beautiful than the early Camaro. The smooth lines are further enhanced by the pleasant lack of chrome tack-on doodads. For the first time the buyer is offered a choice of nose treatment. The standard car comes with a full front bumper slicing across the grill with the turn indicators placed conventionally under the bumper. The Rally Sport, SS, Z-28, etc., models

faced Camaro of old. Our test car also had twin exterior racing type mirrors and the radio antenna built into the windshield; these touches eliminate the wind noise normally generated by these protuberances.

The cockpit, as we said is back three inches from before, and the space has been subtracted from the trunk, small to begin with. Since the wide oval spare tire takes up most of the right side of the trunk, luggage must be carefully chosen and even more carefully packed in order to utilize the available space. On a weekend trip we managed to pack most of our belongings in the trunk, but we used what the British call "soft" luggage, and several small pieces instead of one. With just two people traveling, the back seat area has plenty of room for additional

stuff, and we did use that space to carry the inevitable clothes bag which couldn't fit into the trunk. Years ago a luggage firm manufactured odd shaped luggage just to fit in a Mercedes 300 SL trunk. Perhaps if the pony car trunk space continues to shrink, there could be a ready market for luggage tailored to fit the existing space, side nooks, and corners.

The new Camaro has really wide doors, some five inches wider than before in the current Detroit vogue. Of course these doors make access to the rear seat far handier than in the past. But this is true only if one is climbing rearward in the wide open spaces. If the car is parked diagonally, say in a market lot, it is not possible to get the door open wide enough, even by clunking into the neighboring car's door, to climb in

From the rear view the new Camaro looks like an Italian GT. The smooth lines are classic and only the rumble from the twin exhausts reminds one of its Michigan blood lines.

The small, GT style trunk holds only overnighter size luggage, and soft cases. Side areas are good for flat pieces only, and the husky spare wheel and tire take up 1/3 of the usable space.

come with a split bumperette, a bare grill and indicators installed alongside the single headlight with the look of a real driving light. The main drawback to the latter arrangement is that there is no aesthetically pleasing spot in which to hang the front license plate. The nose piece is fabricated from color matched plastic à la Pontiac and comes to an eagle like point, quite a styling departure from the flat

easily or gracefully. Of course this is a common problem with all wide doored models, but a factor apparently overlooked by the stylists.

The back seat area is tastefully done with reasonable room for children. The normal size adult can ride in the rear for short short distances, but the Camaro is most certainly a 2 plus 2. The rear seats, while adequate in space, are far from comfortable. The front seats have been restyled and are smartly fashionable with small, pop-up style headrests that obviously meet federal requirements but didn't match the head angle of any of our several test drivers. It seems unreasonable in a $5000 car to have relatively uncomfortable bucket seats with no back rake adjustment. The fixed rake does not fit the average physique well, and the addition of reclining mechanisms would certainly help. Leg support is poor from these seats, and these factors combined with the deeply recessed accelerator pedal hung at a steep angle contribute to unnecessary driver fatigue on cross country trips.

The dash panel is all new and faced with nice wood paneling. The instrumentation is a vast improvement from the past with good and easily readable round dials. A big speedometer and tachometer are directly in front of the driver. Oddly enough, even with the RS trim there is only a total mileage counter and no resettable odometer. On the left two small dials read fuel and amperage and on the right matching dials contain a water temperature gauge and a clock. Moving the auxiliary gauges from the console mounting under the dash is a giant forward step for the performance Camaros, but we would like to see an oil pressure gauge somewhere in the line-up. The light switches, a combination windshield wiper and washer switch, cigarette lighter, and slot for rear window defrost switch (not fitted to the test car) balance out the dash. Our SS 396 was complete with a full air conditioning, heating, and ventilation system, so consequently there were no fresh air vents. The mechanical system worked beautifully allowing for enormous varieties of temperatures, and no less than five air outlets festoon the dash. The biggest drawback for the first time driver is the Braille system necessary to work the air controls. The entire panel is placed just over the driver's left knee, and when the steering wheel is in a straight ahead position its broad spoke completely blocks the view to the control panel. Plus the only front seat ashtray is just right of the steering column; and when open it collides neatly with the

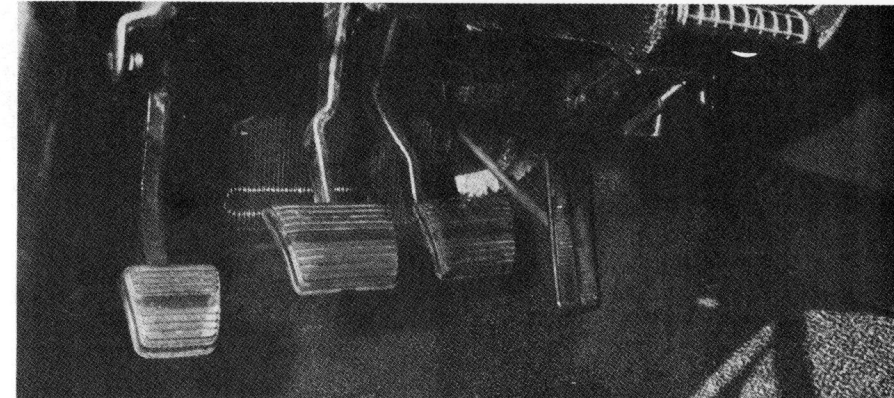

Pedals are reasonably sized, but deeply recessed on the fire wall. The heel and toe technique necessary for starting on a hill is only possible if the driver has a good sized foot.

driver's right knee, and is unhandy for use by the passenger. Since there are no side vents, and the only other ashtray is at the rear of the console, we can conclude that Chevrolet has a subliminal design intended to eliminate smoking in Camaros.

The center console has a handy sized cubby box and there is a tiny glove box on the right side of the dash. Most handy for storage is the open slot just under the radio in the center. Since Camaro and Firebird share a good deal of the parts and pieces, there is an open slot on the Chevy where the stereo tape fits on the Pontiac. The space is quite good for sunglasses, cigarettes, toll change, and the like, and far more accessible to the driver than the stowage spots designed into the car. The stereo tape is not available on the Camaro, and unfortunately, the power windows are not an option this year either. The manual window winders are stiff to use, and the bulky handles look out of place on the slickly styled door.

Interior appointments are completed with a large bright dome light and an under dash light. The test car was fitted with the tilt steering wheel option which was a big help in finding a comfortable driving position. The pedals are all recessed deep on the firewall causing short legged drivers to sit quite close to the steering wheel in order to floor the clutch, and it must be on the floor before a clean shift can be accomplished. The seat belt and shoulder harness are separate pieces of hardware, thankfully offering the driver his choice of restraints.

On the Road

Driving the Camaro is what its all about! The enthusiast can tolerate any number of minor inconveniences if the car has GT manners on the road. The Camaro most definitely qualifies as an American GT, and we think it has the best roadability yet found in a Detroiter. In short, Chevy really has built a Grand Touring car. We are talking about the fully op-

tioned test car of course, but it is the handling SS package that makes the difference. The redone suspension mates beautifully to the big anti-roll bars front and rear, and the wide rim (7 inch) wheels shod with fat, belted, wide ovals. The ride is still a bit harsh by limo standards, but on the boulevard and interstates it is much improved from the old style Z-28 bounce and jounce. The softer spring rates are the big clue here, and there is also less noticable understeer in the Camaro. It is far more docile and neutral in hard cornering excursions.

We had the opportunity to cruise the wonderful Big Sur country on California's Highway 1. The road is famous in sports car circles for being a demanding course of over 50 miles of twisting, winding, tricky road through the cliffs alongside the Pacific ocean. Narrow and extremely rough in spots, the road is made to order for testing a GT. It was here we came to love the SS 396, and found it more than equal to the task at hand. Traffic was extremely light so we were able to extend the car beyond prudence in many cases. The handling package and the limited slip allowed us to corner with gay abandon, and the seemingly unlimited torque of the engine supplied quick throttle steer recovery in almost any gear. It took some practice to avoid wheel-spin when accelerating out of the tightest corners, and we did manage to wear a little rubber off the sides of the wide ovals in our enthusiasm. But overall the SS 396 took to the back road like a real GT, and we think that it is as nifty as most of the higher priced foreigners on this type of road. Contrary to other reports we did not experience brake fade after hard usage in the mountains. In fact we came back raving that

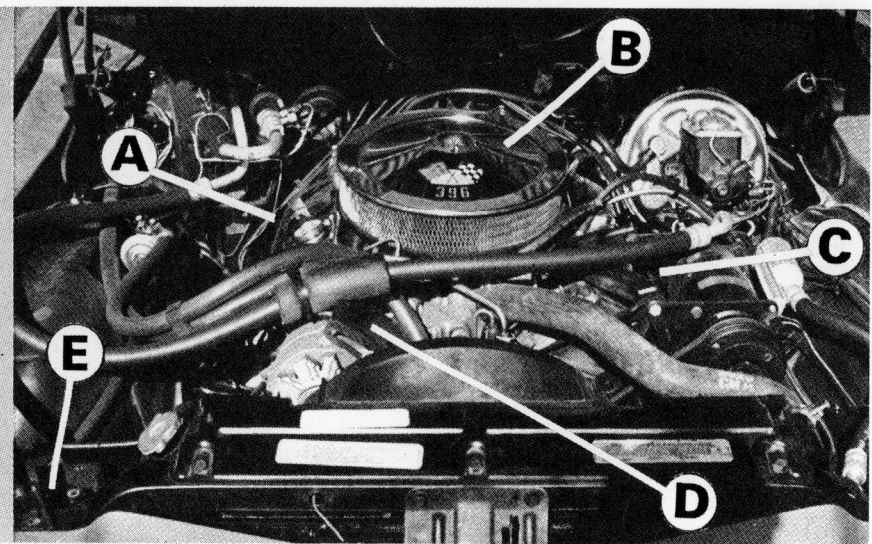

A. Oil filler cap.

B. Low restriction air filter.

C. Spark plug accessibility poor.

D. Radiator, air conditioner and heater hoses clutter under hood area.

E. Difficult access to battery.

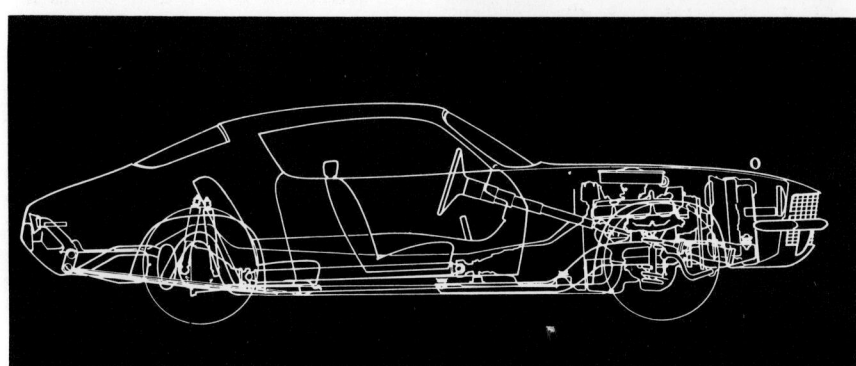

The 396 engine truly fills the bay under the long nose. Chrome trimmed air cleaner and valve covers add the performance touch. Large quantity of hosing supplies the plumbing for anti smog devices and air conditioning.

Diagram shows graphically the long nose and short deck Camaro style. The low roof inhibits the back seat head room and the back seats inhibit the actual trunk space.

Chevy had finally made a car that stopped as well as it went, good for pleasure driving—that is driving for the sheer pleasure of handling the performance in a vehicle on an old-fashioned road.

Of course our SS 396 was a thirsty begger. The California equipped car had a marginal gas tank capacity of 18 gallons, which doesn't take one far if the driver stirs it through the gears. On the highway at cruise we recorded a respectable 13.9 mpg due in part to the axle ratio. In the mountains the figure sank to well under 13, and around town the 396 turned into a gasoholic averaging 10 to 11 mpg. All this juice is the best stuff too, for the SS 396 would pink on lesser brands of hi test. Our test car liked oil also, using about a quart every 400 miles. Of course the SS 396 had a mere 2000 miles on the clock when we reluctantly turned it back to its masters, so it may be that things were just getting seated in the high performance engine at that point.

Summary

The SS 396 Camaro does have the qualities of a true Grand Touring car. Even with the heavy engine, the car handles well on the highways and byways, stops firmly, corners any-where, has more than sufficient power to meet any conceivable situation, and most important, it is fun to drive. The clean and simple styling has definite European tones, in fact the roof line is so low that the tall driver clonks his head entering and exiting just like he would in an Italian GT. Other similarities to Italian stormers make us chuckle a bit too. The lack of stowage space in the cockpit, the ridiculously small trunk (for such a good sized car), the rather minimal back seat comfort and ac-comodation, the blind spots to the rear caused by the fast back styl-ing—it is all part and parcel of a Grand Touring car in the European sense of the term. The Camaro does have fine living space up front, and minor changes could make the seat-ing far more comfortable. Forward vision is excellent, and instrumenta-tion is nearly complete and quite readable. Then too, the car is a re-sponsive machine, makes wonderful noises, causes small boys to whistle, and causes gas station attendants to wash all the windows and ask ques-tions about the performance. Yes folks, we think the 1970 Camaro with SS options is the neatest thing we have ever driven from Detroit's mass production lines. Chevrolet can truth-fully state that they have, at last, built a real American Grand Touring car. ●

Camaro SS396

Data in Brief

DIMENSIONS

Overall length (in.)	188.0
Wheelbase (in.)	108.0
Height (in.)	50.5
Width (in.)	74.4
Tread (front in.)	61.3
Tread (rear in.)	60.0
Fuel tank capacity (gal.)	18.0
Luggage capacity (cu. ft.)	6.5
Turning diameter (ft.)	41.0

ENGINE

Type	V-8
Displacement (cu. in)	402
Horsepower (at rpm)	350 @ 5200
Torque (lb./ft. at rpm)	415 @ 3400

WEIGHT, TIRES, BRAKES

Weight (lb.)	3850
Tires	Firestone W.O., F 70 × 14
Brakes, front	11.0 in. disc
Brakes, rear	9.5 in. drum

SUSPENSION

Front	Unequal length A-arms, coil springs, tube shocks, anti-roll bar
Rear	Live axle, multiple leaf springs, tube shocks, anti-roll bar

PERFORMANCE

Standing ¼ mile (sec.)	15.3
Speed at end of ¼ mile (mph)	92.7
Braking (from 60 mph ft.)	n.a.

The typical Camaro owner is young and enthusiastic about his sporty car, and keeps it in top shape all the time.

CAMARO OWNERS FEEL NO PAIN

Owners reports show performance hath no comfort.

The popularity and fierce owner loyalty associated with imported sports cars a decade ago did not escape the eyes of the hucksters in Detroit. For years American consumer loyalty to the neighborhood domestic dealer had insured a good portion of repeat sales for a given make. Then when huge volume sales dealerships appeared in great numbers in the sixties, prospective customers habitually shopped for a better "deal" rather than a particular brand of car. However, the advent of the personal

car, or the car with a personality, sparked new owner loyalties to individual makes from Detroit. Then the "pony" car mania struck the sales rooms and a new breed of buyer emerged. The "pony" people banded together in groups, some sponsored by dealers and others with no sponsorship, but the sports oriented cars encouraged the cults of Mustangers, Camaro owners, and so forth. Although Ford was first on the market and scored an incredible success with the Mustang, Camaro supporters grew rapidly in numbers over the three-year model span. Today there is a definite rivalry between owners, both organized and individual, of the two makes. Through it all Chevrolet

continues its position as number one in passenger car sales, and the Camaro cult is a big part of that picture by contributing more youth image to the marque rather than the biggest portion of actual unit sales.

The accent is on youth for Camaro promotion. The "hugger" theme of road holding and high performance appeals to the young buyer, and often the Camaro is his first new car. Replies to *Road Test* magazine's owner survey show that 90 percent of the respondents purchased their Camaros new, a very high figure for the sporty car segment of the surveys. The performance image is further enhanced by racing Camaros, both on the drag strips and the road courses of America. The replies showed a great interest in the sporting side of the Camaro, and most owners list the SS or RS options for their car.

Supporting the figures of the survey is our personal observation of Camaro owners. We have seen scores of young people at the local drag strips watching the action and cheering for

The first Camaro in 1967 was an enduring style for three years running. The new body introduced in 1970 is sleeker, lower, but retains the Camaro image of old.

the special Camaros in competition. We also have seen many Camaro owners competing on the quarter mile runs with their own street legal machines. Then too the Camaro is a popular choice for nocturnal street racing. We have also observed flocks of Camaros at road courses around the country, often clustered together in a reserved parking area for Camaros only at some choice viewing point. Chevy power is quite dominant in road racing, and Camaro boosters can be justly proud of their nameplate in road racing. Perhaps the most famous Camaro of all was fielded by Roger Penske for driver Mark Donohue under the racing colors of Sunoco. The dark blue, #6 Z-28 won the Trans-American Sedan Championship in both 1968 and 1969, with the majority of the points-counting race wins going to Mark Donohue. The outstanding success of this particular Camaro was the end result of clever engineering and option writing by the factory, the Penske genius in car preparation, and the remarkable driving talents of Mr. Donohue. This year Camaro road racing fans have another team to cheer on to success. The Penske-Sunoco-Donohue combine is now campaigning Javelins on the Trans-Am circuit, but Jim Hall of Chaparral fame is fielding two sleek new Camaros for himself and Ed Leslie on the 1970 Trans-Am trail. At this writing it is much too early in the season to pick a series winner, but Camaro enthusiasts will undoubtedly line the fences to give moral support for the white Camaros from Texas.

Who are the Camaro Owners

Despite the current trend away from high performance cars, the Camaro people seem to identify with performance, racing, and the sports car mystique. The predominant theme running through the survey lists mentioned the sporty looks, and the fine handling qualities of the car as a reason for buying. Of course, many mentioned marque loyalty as well. The notes often said things like "Always buy Chevy," or "It is a Chevy, what else?" About 50 percent of the replies where multiple car ownership was listed included a Chevrolet sedan or truck in the family garage. This would indicate a strong dealer loyalty perhaps, or we think the out and out preference for Chevrolet over other domestic makes. Many Camaro buyers said that the reasonable price and good "deal" was a big factor in their choice.

The factory figures indicate that over half of Camaro sales are to people 25 years of age and under. The same set of statistics show that 51 percent of Camaro sales are to single men and women, and that one out of every four Camaro buyers is female. According to Chevrolet 9 percent of Camaro

sales are to military personnel, which they claim is the highest percentage of any domestic car. These numbers all match the youth concept of the sales pitch too, and our survey backs up the factory figures on single people, but not on the under 25 age bracket.

The Camaro owner is a cross section of America. Occupations listed ranged from college student through sales, teaching, and engineering to the doctor, lawyer, and similar professions. The average household comes out to three people, although most have either two or four in the immediate family. The majority with more than two in the household also list one or more additional cars; this makes sense when one considers the limited carrying capacity of the Camaro back seat and trunk.

The Camaro owner uses his car

Many Camaro owners are race fans; we see highly optioned units at most road races. Nose styling differences are clear here, but check the driving lights, racing tires, and spoiler on the spectating Camaro on the left. Test unit looks extremely plain in comparison.

more than the average driver. Our figures show that 48 percent drive between 15 and 25 thousand miles per year. A mere 8 percent cover less than five thousand, and 22 percent list 10 to 15 thousand or over 25 thousand as their yearly average of miles covered. Gas mileage figures from owners range from around 10 in town, to over 16 on the road with an overall average of 13 mpg. The gas economy varies greatly with the engine package for sure.

The Camaro buyer in our survey invariably bought his car new, with only 4 percent making a used car purchase. The great majority of buyers listed Mustang and Cougar as other domestics considered before purchase, and 49 percent listed imports, both sports cars and sedans, as other possible cars. The survey indicates that Camaro won out in many cases because of pricing, and

the styling was mentioned in 54 percent of the replies. A degree of owner satisfaction is measurable when 48 percent of Camaro owners surveyed state they would buy another Camaro. Of the 37 percent that would not buy Camaro again, over half listed the SS Chevelle as a probable choice for their next car. These facts go along with the factory theory that Camaro folks stick with Chevrolet when their family outgrows the 2 plus 2 size car. It also supports the premise that performance cars are still quite desirable to a good portion of the market.

Authorized Service

Time was when it was fashionable to ridicule the service supplied by various import makes. In today's world it seems that domestic dealers come in for just as much criticism as the import people. In fact, the domestic dealer service is currently less satisfactory in general than that of the leading imports. The Camaro survey bears out this thought graphically. Some 36 percent of Camaro owners listed the dealer service as poor to shocking, while only 22 percent said service was excellent. In between 13 percent rated dealer service as good, and fair or average accounted for 29 percent. Most of the latter replies had comments that said service was fair for a dealer, but poor by regular garage standards.

Now 62 percent of the survey indicated they were using dealer service, but over half of these replies included the fact that the car was still in warranty. One fellow said that he couldn't wait to get things squared away under warranty so he could move his Camaro service to a better independent garage. The owners quoted exorbitant prices for parts and labor in dealer service, and extreme delays, often days, in retrieving their car from the most routine check-up.

Many stated that the proficiency level of the mechanics was poor, and that they often paid for work that apparently was not done, at least not done properly. Hassling over small items of warranty caused a great deal of comment, and most owners were bewildered by the dealer's reluctance to remedy faults in a car just a few weeks old. The general condemnation of after sales service is not uncommon today, but it certainly seems that improvement of dealer service techniques could be a valuable sales tool for Detroit in these days of declining domestic car sales.

The Camaro owner likes to tinker with his car. Of the 38 percent that do their own service, over half give reasons relating to enjoyment of the work or wanting it done right. Some owners detail with pride their genius in fixing things that were not done by the dealer or factory. Others mention the fact that dealer service is inconvenient or not available geographically. Several answers claimed that Chevy dealers did not stock parts for their car, but that most parts were available on the open market and at more reasonable prices. This was one reason given for doing the work at home or having it done by an independent garage. Overall it would seem that service and warranty problems were the biggest complaint of the Camaro owners in the survey, and one of the features that would keep them from buying another Chevy, or, as several mentioned, any General Motors car.

Likes and Dislikes

The Camaro owner definitely likes the styling and sporty looks of his car. He also likes its performance, and many say that the car is truly a "hugger" just as the ads tell you. By far the most liked feature of the Camaro is its power and good han-

Camaro fans can take pride in this racer fielded by Roger Penske. It won the Trans-Am Championship in 1968 and 1969 with Mark Donohue, pictured here, at the wheel.

dling. Checking through the survey sheets we found frequent comments on good handling, fine disc brakes (an option until '70), and the total fun of driving the Camaro. Now most of these sheets referred to a car with SS and/or RS options, but then the majority of those responding to the survey owned a fully optioned car. Actually we didn't have a single reply from an owner of a six-cylinder Camaro.

Owners like the small size (for a Detroiter) of the Camaro, ease of maintenance regardless of the engine package, and its nimble performance on the road. Many mentioned the comfortable bucket seats as a big plus, but our road testers have never found exceptional comfort in Camaro seats. Of course, owners coming from a typical bench seat would find any bucket more comfortable.

Economy of operation was cited by 52 percent of the survey as a likable feature. Initial cost, gas mileage, and low maintenance needs were hailed as contributory factors to economy of operation. This points up the fact that a goodly portion of Camaro owners have come from the ranks of bigger car owners. Over 70 percent of the owners praised the power disc brakes and many mentioned the safety of the car along with the good handling traits.

Naturally the Camaro owners had their gripes too. Oddly enough, the harsh ride was listed by 50 percent as the most disliked feature. It was the same sheets that listed the good handling and roadability as a plus, so apparently Camaro fans have not accepted the harsher ride as a factor

Camaro folks like to tinker with their favorite car. Many owners feel they can do a better job in service than the authorized dealer, and the owner's survey indicates that a good portion of the week-end mechanics enter competition with their Camaros.

in the optional handling package. The 1970 edition of Camaro has a happier compromise with some suspension changes that offer a softer ride while retaining the fine roadholding capabilities of the SS options.

An overwhelming majority of the replies grumbled about the small size trunk. The trunk space is really tiny for the size of the car, and it is the result of the popular long nose, short rear deck style of the "pony" car in general. The optional small size emergency spare tire helps out in trunk space, but it is still marginal for travel purposes for more than two people. A small percentage listed the auxiliary instrument location on the console as a poor feature. This has been corrected in the current model where the extra gauges flank the speedo and tach directly in front of the driver. About 30 percent of those owners with a four speed manual transmission complained about hard shifting, and several remarked that the Hurst shifter was the cure. Those with automatic transmission had nothing but praise for the three-speed, turbo-hydramatic, and oddly enough, there was no mention of the earlier two-speed automatic.

In general, owners felt their cars were relatively trouble free. Three-quarters of those responding said the car was dust proof and weather proof. Of the 25 percent dissenting, the leaks seemed to be in the trunk area and the windshield, and it seemed most common on 1967 models. There were some complaints on engine performance of cars with full anti-smog devices, but most owners happily reported that proper tuning and careful maintenance of the devices cured most of the symptoms of engine cut-out and the poor cold starting habit. Owners of earlier Camaros said body rattles and shabby quality control in details was typical. Many said that installation of rear mudflaps was necessary to keep the rear flanks clean in everyday operation. The most common mechanical replacement item in the first year of service was the muffler, and, in fact, there was no other replacement part common in the survey. Owners of performance models of the Camaro disliked the rear axle hop and early lock up under hard braking conditions. Some of this has been remedied in the new model with the addition of multiple leaf springs in the rear and other suspension refinements.

Summary

On the whole, Camaro owners are fond of their cars and are true members of the sporty car set. The probable choice of the next car shows the Camaro owner still leaning toward performance. The 48 percent who would buy another Camaro list their satisfaction with the car, its dependability, sporting manners, and the fact that it is a Chevrolet as reasons. The 37 percent who would choose differently next time around indicate other "pony" cars or Detroit muscle cars as the next purchase by 72 percent. The others show a preference for imported cars ranging from Mercedes to Volvo, but invariably a sedan. Of the 15 percent undecided, 47 percent list price and family needs as a big factor in the next car. Others declare that the next one will be an import for greater economy and reasonable size. These folks are generally those with a bitter service complaint with the Chevy. Strangely enough, not one sheet mentioned the high cost of insurance on some of the SS models as a deciding factor either pro or con on the Camaro.

The overall picture of the Camaro owner is still a bit hazy. The biggest part of the survey replies are from owners with highly optioned performance models. So the garden variety Camaro owner is not well represented here. From the survey the Camaro owners are split evenly between single and married people, and those married have a small family averaging one child. Our replies give us an average age around thirty, and a better than average level of income. Over 60 percent of Camaro owners list at least one other car in the garage, and 12 percent own two or more other cars. The Camaro appealed to the owner because of its performance, sporty looks, and competitive pricing. Not one reply mentioned trade-in value as an incentive to buy, so owners apparently will worry about that when the time comes to make the next deal.

The average owner drives many miles a year and he appreciates dependability in his car, while, at the same time, he demands more performance and roadability than the ordinary buyer of domestic machinery. He is willing to put up with dealer service in order to get warranty satisfaction, but he is convinced that general service is better and cheaper on the outside. The typical Camaro driver is a real enthusiast who likes to drive hard enough to appreciate good handling in a car, and he is pleased with the racing success and competitive image of the Camaro. A good many Camaro owners participate in an occasional rally or slalom and have a measure of success in grass roots competition.

Until the 1970 Camaro debuted last winter there had been few changes in the car since introduction in late 1966. It will be most interesting to see the results of the survey a year from now. The new model with its different body and a more European aspect has many changes, most for the good. The optioned Camaro handles better, rides a good deal better, and the entire cockpit layout is much improved. However, it is heavier, longer, and the smaller V-8s are no longer available. As with everything these days the base price is up a bit, and the overall fuel economy is down. Still, we predict the new Camaro owner will continue to choose the car for its looks and fine road manners. We are curious to see how the owner's likes and dislikes agree with the staff evaluation of the Camaro in this issue, but it will have to wait until next year. ●

YENKO'S "SUPER Z" 400-inch CAMARO

Is this the 1972 Z-28?

By TERRY COOK

It looks as if another 50 cubic inches of displacement has been added beneath the hood of your favorite type of Camaro. Although we're not speaking of a production line option, the 400 cubic inch Camaro will be available through Yenko Sportscars, Inc., in Canonsburg, Pennsylvania, as a second cousin to the real Z/28.

Now if the whole idea of a 400 cubic inch small-block Yenko Camaro sounds strangely familiar to you, perhaps that's because we already did a story on this package way back in the October, 1969, issue of *Car Craft* (Teach Your Old Dog New Tricks). Of course, at that point Yenko wasn't involved, but we understand they are currently taking orders for 400-inch Camaros.

The beauty of the car is that it may well qualify as a regular low insurance vehicle, rather than being classified as a supercar. This may come to pass because the 3200 pound car will be rated at about 290 horsepower (the regular two-barrel equipped 400-inch

(continued on page 121)

Low end torque of new "Super Z" from Yenko was demonstrated at the proving grounds; rpm potential, however, is limited despite hot hydraulic cam. The car's horsepower (290) to weight (3200) ratio should make it a low insurance package.

SPECIAL CHEVROLET ISSUE

RAC TITLE CHASER

EXTERIOR PHOTOGRAPHS of the Wiggins Teape backed Camaro accurately reflect some of the £8,000 spent last year in keeping it in the style to which it had become accustomed. The detail photographs show the 5-litre Z28-based V8 engine, minus exhausts in readiness for a replacement Chevrolet 5.7-litre engine, the leaf-spring suspension and Panhard rod arrangement at the rear, and the contrastingly small bucket seat, surrounded by roll cage, which Australian Brian Muir occupies during racing hours. The GM ventilated disc brakes, fitted on all four hubs, now have excellent Lockheed calipers following a season of development in which the clutch had to be changed and in which two engines were consumed.

TO SCORE an outright win in saloon-car racing days you need an American "Pony Car" like the one shown here and, most important, a great deal of money to support its sub-5-m.p.g. fuel consumption and general thirst for items such as tyres. Former GLTL Lotus 62 driver Brian Muir has conducted this particular example since it was purchased in February 1970 by his building contractor entrant, Malcolm Gartlan. Financial backing to the total of £9,000 for 1971 comes from the papermakers Wiggins Teape, without whom the Chevrolet could not run in the poorly rewarded RAC Saloon Car Championship. Last year Gartlan reckoned to have spent at least £8,000, including a £1,600 tyre bill incurred by the temporary need for imported Trans-Am tyres: the result of this expenditure was a fourth overall in the RAC series and second in class behind the ex-works Boss Mustang of Frank Gardner, plus several outright race wins.

This Chevrolet Camaro was bought from German Peter Reinhart after he had returned from his previous job with Roger Penske Racing in the USA. The car was originally intended as a back-up vehicle for Penske's successful 1968 attempt on the SCCA Trans-Am title, but as it was not needed Reinhart completed the preparation in Germany, and competed promisingly on the Continent before Gartlan bought it on the recommendation of his chief mechanic—Ted Grace. The latter is assisted in the modern Cotswold workshop by Patrick Salter.

The Z28 tag refers to the car's original specification of engine and transmission items specially homologated for racing: however, for next year the label will no longer be applied as the 5-litre Z28 engine, built up by Grace after two previous engines had blown up (including the well-used 418-b.h.p. Traco unit), is to be replaced by a 350 cu. in. (5,740 c.c.) Chevrolet V8, if all goes well. Incidentally, our picture of the engine clearly shows the staggered inlet arrangement of the twin, four-choke, Holley carburetters which do such an effective job in supplying fuel for many American racing V8s.

The four-speed gearbox was manufactured by Muncie, a General Motors subsidiary, and it transmits power *via* a 10-in. plate Borg & Beck clutch of the type used on F5000 cars. Five final-drive ratios can be chosen from, varying from 4-to-8.1 to 3-to-1. Approximate gear speeds on a fast track would be 80 m.p.h. in 1st, 105 m.p.h. in 2nd, 140 in 3rd, and as much as 165 m.p.h. in 4th if the driver is brave enough to leave all four 11¾-in. GM disc brakes alone in 28 cwt. of machinery.

Suspension features many GM optionally available parts for the unequal-length wishbone front and leaf-sprung rear end with its "one either side" staggered shock-absorbers, Panhard rod and two locating radius arms. Minilite wheels of 10-in. rim section are used in conjunction with Firestone 14 front tyres and the same company's "slick"-style rears.

Altogether an immaculately prepared car which keeps the spectators entertained in the hectic 20 minutes or so of saloon-car racing which usually follows major British meetings these days.—J. W.

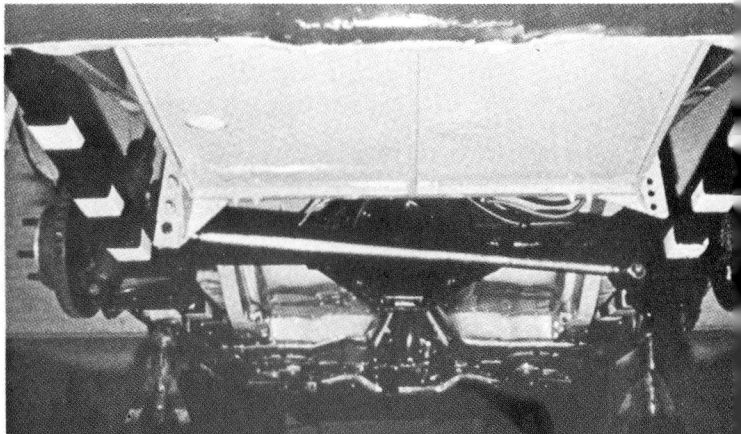

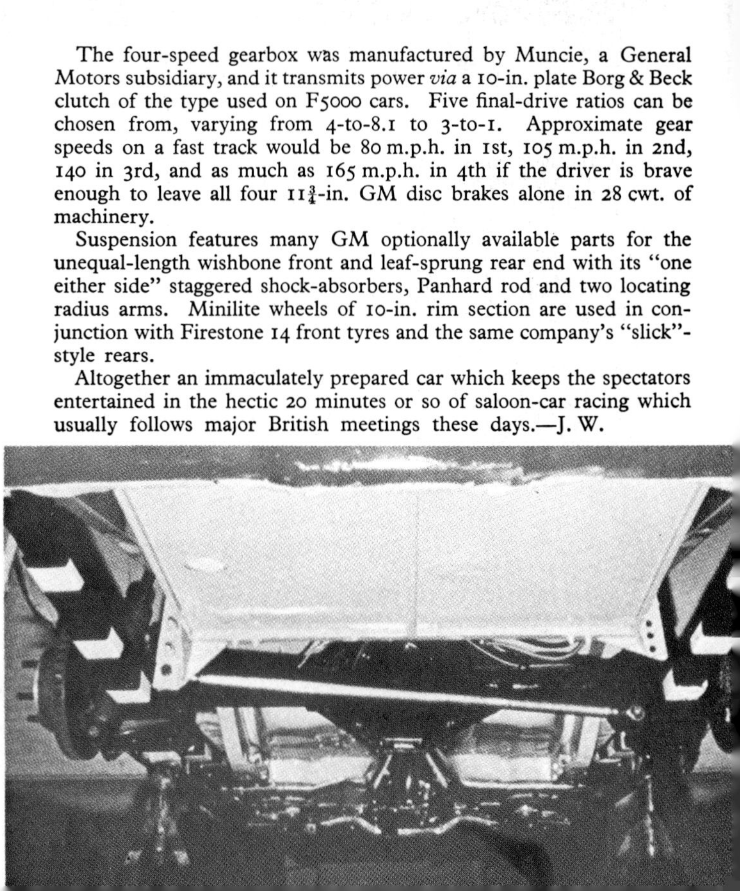

CHEVROLET
CAMARO

The present Camaro must be compared with the kind sold up until April of 1970, as the '71 models are a carryover of the post-April offerings. The only appearance changes are some new wheel covers, a redesigned taillight lens with a larger back-up light, two new colors for the optional vinyl roof and the availability of full-width front and rear spoilers with any of the several engine options.

Compared with the original, the present body style is 2 inches longer on the same 108-inch wheelbase, plus being lower and wider. The body shell in effect was moved 3 inches to the rear without stealing anything but a little rear seat headroom from the occupants. In fact, most key interior dimensions were increased fractionally. The shift was motivated primarily by styling considerations that were, in turn, dictated by the decision to offer only a single body style of the fastback type. In the past, the hardtop roof shape was formed around the common usage with the convertible of the same rear quarter panels and deck lid.

CAMARO Z-28

CAMARO COUPE

CAMARO SS

Protective features incorporated into the new (last April) body include a metal barrier between the passenger and luggage compartments, door beams for side impact protection and energy absorbing windshield pillar moldings. The roof design uses two panels, the inner one forming box sections for the windshield header, side rails and rear header. In addition, the stub frame on the front of this otherwise unitized structure was redesigned for greater strength. Suspension changes made at the time included staggered rear shock absorbers for all cars and not just the performance versions, a wider front tread and an optional handling package that uses the regular springs to retain some measure of ride softness combined with front and rear sway bars and stiffer shocks for the desired extra stability. Front disc brakes also became standard equipment as well.

Engine and transmission combinations are continued as before except for some whopping cuts in horsepower for the SS and Z-28 models. The 402-cubic-inch SS option, for example, loses 75 (from 375 to 300) and the Z-28 goes from 360 to 330. Restoring these cuts, if you're so minded, won't be easy because in most instances compression ratio was lowered by dishing the pistons. Also cams, distributor timing and carburetor calibrations are different.

The base engine is the ubiquitous 250-cubic-inch, 7-main-bearing 6 newly rated at 145 hp. You can get this with either a 3-speed manual or 2-speed Powerglide automatic. The standard V-8 is a 307 2-barrel which retains its 200-hp rating and adds the desirable 3-speed Turbo Hydra-Matic to the transmission list. After this, the 3-speed manual box and Powerglide disappear, a floor-shifted 4-speed becoming standard and Turbo Hydra-matic optional. There are 2-barrel and 4-barrel versions of the 350 with 245 and 270 hp, the Z-28 V-8 already mentioned, and the 402 SS option is also available.

As before, RS is a styling option with a body-colored Endura grille wraparound, split front bumper and parking lights adjacent to the headlights that are styled to look like halodide road lights. The SS, which can be ordered alone or in combination with RS, includes one or the other of the big engines, power brakes, 14x7 wheels and sport suspension with the bigger of the two engines. The Z-28, which also can be mixed with RS, uses 15-inch wheels, a heavy-duty radiator and special springs along with front and rear stabilizers. You also get an instrumentation package which is now up where you can see it and not on the console, front and rear spoilers and a blacked-out grille. Concealed windshield wipers come with any of these packages and are also technically an option by themselves, but we have yet to see a Camaro pictured without them.

CAMARO RALLY SPORT

Bucket seats with integral head restraints, formerly an extra-cost item, are now standard and some of the switches are bestowed with those European-style squigglings to indicate their function. Also, there's an optional 4-spoke sport steering wheel offered. The rear seat is strictly for two and the fold-down option is no longer available.

Despite the single body style compared to the Mustang's three, Camaro forged into sales leadership among pony cars shortly after its mid-year introduction. How it will fare against the all-new '71 Mustang line, with three bodies, remains to be seen. •

CAMARO/SS

ENGINES: 250 cu ins (145 hp). 307 cu ins (200 hp). 350 cu ins (245, 330 hp). 402 cu ins (300 hp).
TRANSMISSIONS: 3-spd std (250 and 307), 4-spd manual std (350 and 402). 2-spd auto opt (250 and 307), 3-spd auto opt (all but 250).
SUSPENSION: Coil front, leaf rear.
STEERING: Manual std, variable-ratio power opt, curb-to-curb 39.0 ft.
BRAKES: Front discs std, power opt.
FUEL CAPACITY: 18.0 gals.
DIMENSIONS: Wheelbase 108.0 ins. **Track** 61.3 ins front, 60.0 ins rear. **Width** 74.4 ins. **Length** 188.0 ins. **Height** 50.5 ins. **Weight** 3208-3320 lbs. **Trunk** 7.3 cu ft.
BODY STYLE: 2-dr hdtp.

ACCELERATION standing ¼ mile, seconds

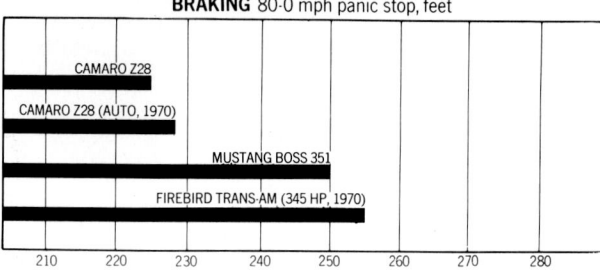

CAMARO Z28	
CAMARO Z28 (AUTO, 1970)	
MUSTANG BOSS 351	
FIREBIRD TRANS-AM (345 HP, 1970)	

Scale: 13 14 15 16 17 18 19 20

BRAKING 80-0 mph panic stop, feet

CAMARO Z28	
CAMARO Z28 (AUTO, 1970)	
MUSTANG BOSS 351	
FIREBIRD TRANS-AM (345 HP, 1970)	

Scale: 210 220 230 240 250 260 270 280

FUEL ECONOMY RANGE mpg

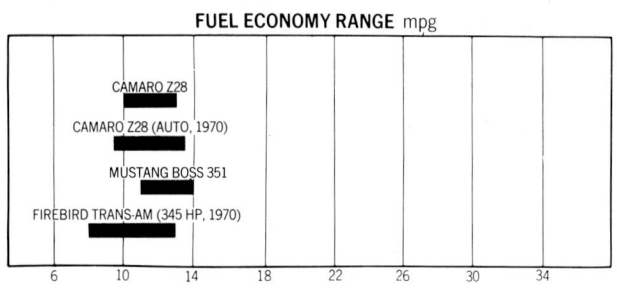

CAMARO Z28	
CAMARO Z28 (AUTO, 1970)	
MUSTANG BOSS 351	
FIREBIRD TRANS-AM (345 HP, 1970)	

Scale: 6 10 14 18 22 26 30 34

PRICE AS TESTED dollars x 1000

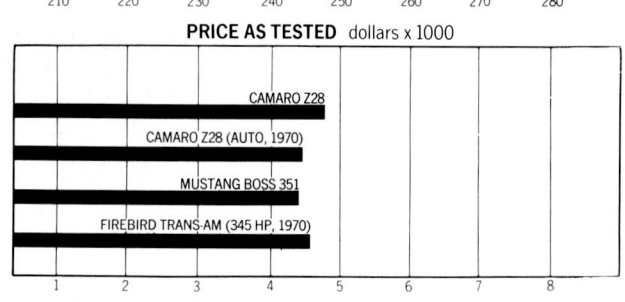

CAMARO Z28	
CAMARO Z28 (AUTO, 1970)	
MUSTANG BOSS 351	
FIREBIRD TRANS-AM (345 HP, 1970)	

Scale: 1 2 3 4 5 6 7 8

Chevrolet Camaro Z28

Manufacturer: Chevrolet Motor Division
General Motors Corporation
Detroit, Michigan 48202

Vehicle type: Front engine, rear-wheel-drive, 4-passenger coupe

Price as tested: $4760.65
(Manufacturer's suggested retail price, including all options listed below, Federal excise tax, dealer preparation and delivery charges, does not include state and local taxes, license or freight charges)

Options on test car: Base Camaro V-8, $3011.00; Rally sport, $179.05; Z28 package, $786.75; 4-speed transmission, $205.95; power steering, $115.90; deluxe seat belts, $15.30; console, $59.00; tinted glass, $40.05; auxiliary lighting, $15.80; AM/FM, $139.05; rear speaker, $15.80; tilt steering wheel, $45.30; sport steering wheel, $15.80; custom interior, $115.90.

ENGINE
Type: V-8, water-cooled, cast iron block and heads, 5 main bearings
Bore x stroke . . . . 4.00 x 3.48 in, 101.6 x 88.4 mm
Displacement . . . . 350 cu in, 5740 cc
Compression ratio . . . . 9.0 to one
Carburetion . . . . 1 x 4-bbl Holley
Valve gear . . . Pushrod operated overhead valves, mechanical lifters
Power (SAE) . . . . 330 bhp @ 5600 rpm
Torque (SAE) . . . . 360 lbs/ft @ 4000 rpm

Specific power output . . . . 0.94 bhp/cu in, 57.5 bhp/liter
Max recommended engine speed . . . . 6500 rpm

DRIVE TRAIN
Transmission . . . . 4-speed, all-synchro
Final drive ratio . . . . 3.73 to one

Gear	Ratio	Mph/1000rpm	Max. test speed
I	2.20	9.1	59 mph (6500 rpm)
II	1.64	12.2	79 mph (6500 rpm)
III	1.27	15.8	103 mph (6500 rpm)
IV	1.00	20.0	104 mph (5200 rpm)

DIMENSIONS AND CAPACITIES
Wheelbase . . . . 108.0 in
Track, F/R . . . . 61.3/60.0 in
Length . . . . 188.0 in
Width . . . . 74.4 in
Height . . . . 49.1 in
Ground clearance . . . . 4.2 in
Curb weight . . . . 3560 lbs
Weight distribution, F/R 56.2/43.8%
Battery capacity . . . . 12 volts, 61 amp/hr
Alternator capacity . . . . 444 watts
Fuel capacity . . . . 17.0 gal
Oil capacity . . . . 4.0 qts
Water capacity . . . . 16.0 qts

SUSPENSION
F: Ind., unequal length control arms, coil springs, anti-sway bar
R: Rigid axle, semi-elliptic leaf spring, anti-sway bar

STEERING
Type . . . . Variable ratio recirculating ball, power assist
Turns lock-to-lock 2.3
Turning circle curb-to-curb . . . . 39.0 ft

BRAKES
F: . . . . 11.0-in dia. vented disc, power assist
R: . . . . 9.5 x 2.0-in cast iron drum, power assist

WHEELS AND TIRES
Wheel size . . . . 15 x 7.0-in
Wheel type . . . . Styled, stamped steel, 5-bolt
Tire make and size . . . . Goodyear F60-15
Tire type . . . . Bias-belted, tubeless
Test inflation pressures, F/R . . . . 24/24 psi
Tire load rating . . . . 1500 lbs per tire @ 32 psi

PERFORMANCE
Zero to	Seconds
30 mph	2.5
40 mph	3.6
50 mph	4.9
60 mph	6.7
70 mph	8.7
80 mph	11.1
90 mph	13.8
100 mph	17.1

Standing 1/4-mile . . . . 15.1 sec @ 94.5 mph
Top speed (at redline) . . . . 130 mph
80-0 mph . . . . 225 ft (0.94 G)
Fuel mileage . . . . 10-13 mpg on premium fuel
Cruising range . . . . 170-221 mi

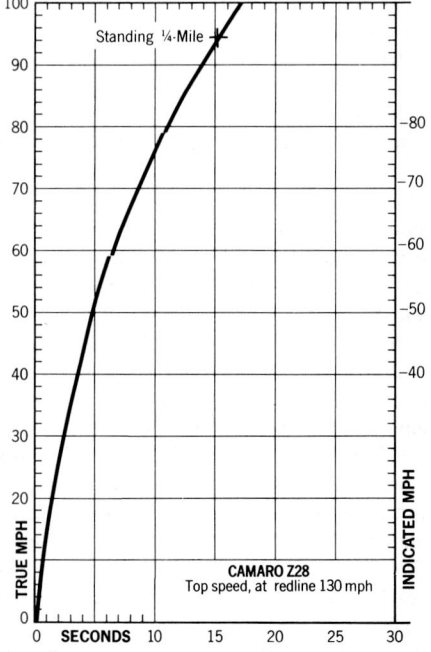

CAMARO Z28
Top speed, at redline 130 mph

CHEVROLET CAMARO Z28

Underneath last year's smooth exterior beats 1971's low-compression engine.

Chevrolet, by virtue of its own diligence and a generous helping of market savvy, has worked itself up to the position of first in line for second fiddle with the Camaro. First in line because it appears that the Camaro is finally going to outsell the Mustang in the sporty car market, but second fiddle because the sporty car market itself is withering like last week's roses—it's not nearly as important as it used to be. In 1967, the best year ever for sporty cars, they captured 11.9% of the total car market. Since then it's been all down hill to the point where only 7.2% of 1970 auto sales were sporty cars.

Traditionally, the Mustang has outsold the Camaro on a three-to-two basis. That has changed now, probably due to the wind-swept styling and the engineering sophistication introduced on the 1970½ model, so that as this is being written the Mustang and Camaro are neck and neck in the market. And the shock waves from the GM strike haven't all been damped yet. If the *C/D* Readers' Choice Poll is any indication (the Camaro received twice as many votes as the Mustang), the Camaro should pull into a clear lead this year.

If this does happen, it will be contrary to Detroit's game plan. The Mustang should be this year's high fashion. Ford spent millions to restyle it for 1971 while the Camaro continues almost unchanged. And to Ford's discomfort, the Camaro seems better suited to the market. Its styling is functional, honest rather than tricky, and to the driver the Camaro feels compact and agile—qualities which we feel are prerequisites for any sporty car and which are unfortunately absent in the Mustang.

Along with *C/D* readers, Chevrolet is obviously not displeased with the Camaro. For 1971 there have been few changes—only two major ones, in fact—but they are conspicuous enough to alter the personality of the car.

Of the major changes, the only visible one is the front bucket seats. They are of a high-back design now, incorporating the head restraints into the seat back, and are made by a new manufacturing process which provides very deep foam cushions. Your backside will know the difference between a new Camaro and last year's model in a millisecond. The foam welcomes you in like quicksand and the first impression is of bottomless padding. But that changes after a few miles. Then you start squirming around,

PHOTOGRAPHY: HUMPHREY SUTTON

trying to find a position that is comfortable, one that doesn't make you feel awkward behind the wheel, and it is not to be found. The back rest is too vertical, not enough angle between it and the cushion, so that the seat feels like a half folded-up lawn chair. A two-position backrest, like the Vega's, or a recliner mechanism like most bucket-seat imports, is a logical addition and it is a measure of Chevrolet's nearsightedness that one is not available in an otherwise well-planned car.

The idea of foamy-soft seats in a driver's car like the Z28 needs to be reviewed also. They cause no problems in normal transit but when you start to *drive*, move the controls firmly and quickly, you find that you do it from a moving platform. When you depress the clutch you sink further back into the seat and when you notch the shift lever through a detent you find that your body moves almost an equal distance in the opposite direction. And tightening the seat belt until your legs turn blue doesn't cure the problem. You end up bracing against the steering wheel. We think the solution is clear. There is no substitute for a firm, well-shaped seat in a car where the driver will be active. Monte Carlos are another story.

119

Z28

Those who have been attracted to the Camaro by its reputation for performance will view the other major change, one you can neither see nor sit on, as an insidious and heretical departure from the straight and narrow. What they did, those Chevrolet engineers, in political lip service to the advantages of low-lead gasoline, was to lower the compression ratios of all performance engines to 8.5-to-one. The only exception to that policy is the Z28 which dropped to 9.0 from 11.0-to-one last year. That move is particularly detrimental to an engine which relied on high specific output rather than enormous displacement for its energy. And now we are to the central reason for this test. How much performance does the Z28 lose in its transformation. It turns out that it is not an easy question to answer and a one-car evaluation is not conclusive. Chevrolet engineers say that, statistically, the compression ratio reduction drops the torque curve by about 5% throughout the range and weakens power output by about 15 horsepower—to a net value of 275 hp for the 1971 Z28. It's advertised at 330 hp, down 30 hp from last year and it still requires premium fuel to operate without discomfort.

After explaining all of this, the engineers hasten to add that there can be more than a 15 hp variation between two engines built on the same day so that it will not be unusual to find 1971 Z28s which are faster than the 1970 models. Then, too, there will be an enormous gap between a strong 1970 model and a weak 1971. Somewhere in this maze of production tolerances lies the 1971 test car and the Z28 we drove a year ago (May, 1970). The one last year, an automatic with a 4.10 axle ratio, was plenty quick—14.2 seconds at 100.3 mph in the quarter. Its successor, a close ratio 4-speed with the standard 3.73 axle ratio, is a 97-lb.

weakling by contrast—15.1 seconds at 94.5 mph on the dragstrip. There is far more than 15 horsepower between them. The new model was free of conspicuous ailments—it would readily pull the 6500 rpm redline in the first three gears (the test track wasn't long enough for fourth)—it just wasn't very powerful, particularly not at low speeds. Sidestepping the clutch at 3000 rpm would produce only a few feet of wheelspin on Orange County International Raceway's asphalt. Then the tires would key in and the engine would sag until it could climb back up on the torque curve. Here is where the wide-ratio (2.52 low gear instead of 2.20) transmission would have been advantageous. Curiously, most Z28s are ordered with the close-ratio 4-speed because customers think it's the hot set-up. It is for road racing but any car that has to stop and start, which includes all street cars and all drag racers, will benefit from the extra first gear reduction of the wide-ratio gear box.

As for the test car's performance, there are no excuses to be made. The car was surely within production tolerances—it was just on the low side, a car that should get its ultimate owner a small rebate on his insurance policy's muscle car surcharge.

Most of the car-to-car variations in driveability have their origins in the emission control modifications. As one Chevy engineer put it, "All of our engines will respond to more spark (advanced ignition timing) and more fuel because we've got them leaned out so far for emissions." Still, maximum performance has not been seriously reduced because wide-open throttle operation of a powerful engine like a Z28 is not a situation that falls under the scrutiny of the emission control test. Rather it is the normal driving modes—cold start, warm up, throttle response and smoothness under part-throttle operation—that have become unpredictable. The Z28's full-throttle mixture has not been changed for 1971 although part-throttle has been leaned out. And while the Combined Emission Control valve. eliminates vacuum advance in first and second gears on both manual and automatic Z28s, it really has no effect on performance because there is no vacuum advance at wide-open throttle anyway. Part-throttle driveability definitely suffers, however, and the test car was appropriately spastic, particularly before it was fully warmed up. Which was not entirely unexpected. Automatic transmission-equipped cars are more tolerant to lean mixtures than manuals so, no matter how carefully the factory assembles the parts, there are bound to be more balky 4-speed cars.

The CEC valve mentioned earlier also has another job—that of holding the throttle partially open during deceleration (which aids in hydrocarbon control). The valve is activated in both third and fourth gear on a 4-speed and in third gear on an automatic. It may do good things for smog but not for the driver. Engine braking is about as effective as dragging your foot on the pavement. And of course the valve's adjustment is subject to the same production variations as apply to the carburetor.

The sad part is that, as a customer, you pay your money and take your chances. If you get a good one, a Z28 where all of the tolerances are optimum, it will run as strong and be nearly as smooth and responsive as an uncontrolled car. But if you get a bad one, one that starts poorly, runs roughly and has poor throttle response—and this applies to all cars, particularly those with manual transmissions, not just the Z28—you're stuck. The system is so complicated with sealed mixture adjuster screws, time relays, cold overrides, external valves and internal jets so fine that the naked eye can't distinguish between a good one and a bad one that it is way beyond the understanding of most dealership mechanics. Even if they could figure it out they wouldn't have the sophisticated test equipment to isolate the problem. Already there are complaints of service managers throwing up their hands, blaming the emission control system for all manner of problems and swearing that nothing can be done.

As for Z28 owners, if your new car doesn't drive as well as you think it should, don't despair immediately. Engineers say that Z28s smooth up after they are broken in—about 1200-2500 miles depending

upon how the car is driven. The nature of the camshaft is such that low speed output is weak, sometimes barely enough to overcome the pre-break in friction, so that idle and off-idle of a "green" engine will be unsteady at best. And all of this can be aggrevated by the mechanical lifters. It is not unusual for engines to leave the factory with too little lash, a situation which also contributes to roughness and reduced power. Other than the problems related to the camshaft, engineers say that the Z28 has no special emission control difficulties. The large intake manifold runners, ports and valves, while they do slow down air flow velocities, are less significant than the camshaft.

All of this means that the days of the solid-lifter Z28 that we've all known and loved are numbered. It will survive this year and maybe next, but it will never make 1975. Fortunately, in packaging the engine, Chevrolet has learned a great deal, about building successful sporty cars. The Z28 is very definitely a driver's car, quiet and comfortable except for the upright seat, with extremely quick power steering, accurate short-throw shifter and powerful brakes. Only the Pontiac Firebird Trans-Am, a sister under the skin, is in the same league. The two of them are clearly the standouts of the sporty car field—the only two that haven't lost sight of the mood. GM stylists must be a benevolent sort. Instead of marauding over the project, drawing lines and molding shapes that obscure the car's function, they have released a car that is almost understated by Detroit standards. The virtue of that will be more obvious as time passes; when today's Camaros still look attractive on tomorrow's freeways.

As a racer, however, it would appear to be all over for the Z28, at least as far as the Trans-Am is concerned. There will be no factory cars this year, under the table or otherwise, and an independent, no matter how good he is, won't have a chance against the Penske Javelin. But for those who like to pretend, the huge Ferrari Dino-style rear spoiler offered last year at Jim Hall's insistence is still available as an option along with a not-so-effective front air dam. Only a small rear ducktail is standard equipment with the Z28 package—that and a reputation.

For 1971 the reputation will get no help from the race track and the street models have been sadly undermined by the anti-emission engineers. But even so, the Camaro still looks like more than a match for the Mustang. ●

engine is only rated at 250 hp) and will therefore not be on the supercar side of the 10 horsepower/pound yardstick used by many insurance companies.

If you're at all familiar with the 400 cubic inch small-block Chevy engine, you know that it isn't the greatest performance combination going. With connecting rods that are .138-inch shorter than the rods used in all other Chevy small-blocks, a nodular cast-iron crank, low performance cylinder heads and a set of heavy low compression pistons, it seems that the engine has a number of strikes against it. However, with the addition of a few existing components such as the aluminum single four-barrel intake manifold off the existing Z/28, an 800 cfm Holley four-barrel carb, and the hot hydraulic cam, similar to those used in the 350/327 Chevy II's, the engine really comes alive. While the compression has been reduced from the Z/28's previous 10½:1 down to 8½:1, the combination works nicely, even on regular fuel. The warmed over 400 incher may not have the top end of some Z/28's of the past, but the Yenko combination has gobs of low end torque to make it a nice street machine.

Remember when Chevy's first Z/28 was 302 inches, and just about the time Ford came out with their Boss 302 the Z/28 jumped to 350 inches with the LT-1 engine? Well, now that Ford is releasing their Boss 351 to catch up to Chevy in the displacement race, it looks as if the Yenko Camaro has hopped another 50 cubic inches ahead with the 400. Who knows, if enough of you like the idea of a 400-inch small-block, low insurance Camaro, Chevrolet might even make it into a production line type Z/28 for 1972. ©

THE LAST ROUNDUP

Comparison

It's an early John Wayne flick, maybe even pre-war. The Cattlemen vs. the sheepmen. The cattle guys are heading their Texas longhorns in a big drive up to Abilene or wherever, and, suddenly, there are barbed wire fences on the old Chisholm Trail. With guys sitting by the fences with 12-gauges on their knees. "Well," says Duke, "it looks like they ain't a-gonna let us run no more big drives. Gotta send 'em on the railroad." Of course, they never got this far without guns a-blazin' but that's what he *would've* said if he could've finished the sentence.

In a way, that's what is happening in Detroit with the ponycars. Oh sure, they were the marketing guys' baby once, starting in April, 1964, with a little ol' sure-footed cayuse called the Mustang. It was a spritely lil' devil, with a six-cylinder under the hood, a tiny back seat and a rock-bottom price tag of only $2,368. Ford sold over 248,000 Mustangs the first year. The next year, 1965, they sold half a million and duplicated this figure in '66. But,

by then, things were already changing. Over at the Chrysler spread, they had put a fastback on the Valiant and named it after a vicious fighting fish. At the Chevy spread, they were combing the parts shelves for a sportster named after a Latin word they made up. And even lil' ol' AMC was getting it on, sinking their Marlin to come up with the Javelin. Soon the great big old pie was cut up into so many little pieces the market was saturated.

Then the Boom Town went bust. In 1967, Ford watched its high-water figures grow anemic as they dropped over 160,000 in sales from the year before with their Mustang. Now, according to 1970 statistics, Ford's Mustang sales are less than 165,000 units per year, below what they were back in the beginning, in 1964.

Of course, it wasn't really the competition that killed the ponycar. It was the auto makers themselves. You see, selling cars per se is not really a very profitable enterprise. It's what you sell *with* the car that makes the whole venture worthwhile. If you can take, say, a $2,500 Plain Jane Mustang and talk the buyer into loading it up with power steering, power brakes, a vinyl roof, air conditioning, AM-FM stereo, mag

wheels, a luggage rack, stripes, scoops, and a big V8 to pull it all, then, my friend, you've got a *sale*. In that $1,000 worth of options is your *crème de la creme* — the *profit*. That's why the original six-cylinder Mustang sat at the back of the showroom while the salesmen pushed the optional V8, first a 260 and then a 289. The escalation derby gathered momentum as time went on, until by the late sixties, you could buy Mustangs with up to 429 cubic inches of V8 — engines that originally had been developed for the 200-mph straightaways of Daytona. And the prices of these overbuilt ponycars rose pretty well out of sight, too — in the $5,000 bracket rubbing elbows with Buicks and T-Birds. Option overkill.

Where will it all end? We figure the ponycars will all fade out into the sunset by 1975, along with that wild girl from Laramie. Oh, at least one auto maker might keep his ponycar nameplate alive in his line as a sort of nostalgic reminder of When Things Were Good. But there are all sorts of tides running against the ponycar now — things like bumper laws that will require your 1975 Camaro to take a 5-mph cuff on the chin with no damage to the body. And then there's the Muskie Bill,

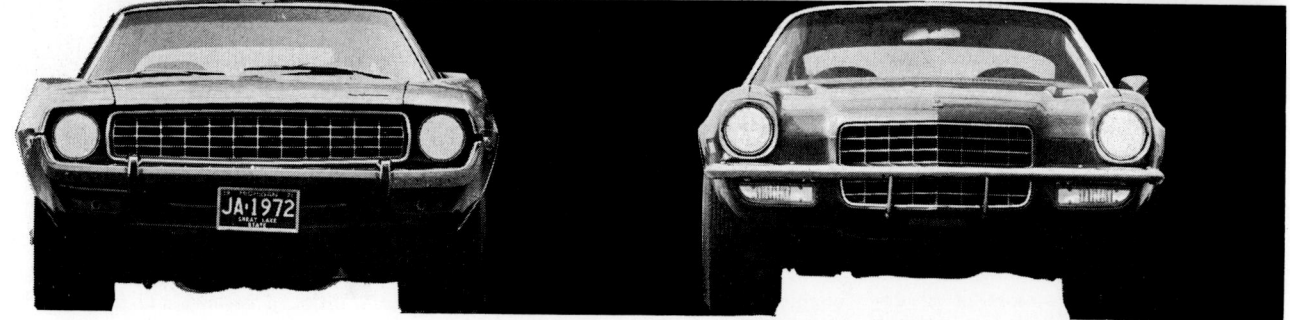

Whatever happened
to the ponycars?
Oh, they're still here.
The only trouble is,
the ponies are now
horses and they
cost $5,000

which pretty well emasculates the high-performance engines.

CAMARO

The Camaro interior is like the exterior — tidy. Our only complaint would be that the accessory gauges in the Special Instrument package are a bit tiny — hardly more than 2 inches across. But, as if to make up for it, the tach is huge, just like the Corvette's.

The Camaro buckets are cushiony enough and tasteful with their no-sweat cloth inserts, with a bit more lateral support than the Mustang seats.

Our Camaro was equipped with the 350-cu.-in. V8 "economy" engine, running a Rochester two-barrel. Last year, this engine put out 245 gross horsepower at 4800 rpm, and is the same this year. With its low 8.5 to 1 compression ratio, it'll burn regular without pinging. The trans in our test car was GM's famed Turbo Hydra-Matic, a three-speed automatic, with a floor-mounted shifter. You can shift manually with

this setup but, as with the Mustang 302 two-barrel, Chevy's 350 two-barrel engine doesn't have that great a power range to want to take advantage of this feature at stoplights.

In straight-line acceleration, the Camaro was the toad of the pack, edging through the eyes at a top e.t. of 18.4 seconds — almost a second slower than the smaller-engined Mustang. Perhaps it's unfair to comment on a car not set up for drag racing, but the performance of the Camaro grew worse with each successive drag strip run, indicating

>>>>

ROUNDUP

that, at least for racing, the car needs lots of cooling and fuel feed aids.

The cornering of our test Camaro was adequate — with a fair amount of plow, but somehow even the variable-ratio power steering didn't seem as deft as the Mustang's steering. Our Camaro boasted a front anti-roll, or sway bar but for the rear bar, we would've had to order the LS3 402 engine, the Z/28 or F41 h-d suspension. Since we ran only E78 treads, instead of the F70s like the SS or Z/28's F60s, the bar wouldn't have made much difference.

FIREBIRD

For our Firebird we were delivered an Esprit, the third one down on Firebird's pecking order — sort of a detuned hot dog fitted with a two-barrel

carb for pinching pennies on fuel costs. We suspect this one is aimed at secretarial types who want style and a little oomph under the hood but still want to save on gas. The 400-cu.-in. 2v engine was rated at 265 gross horsepower last year but this year you'll have to get used to an anemic-sounding but more realistic 175 *net* hp rating. Options on our test car included GM's famed three-speed Turbo Hydra-matic with a floor-mounted shifter.

One word about the Endura bumpers up front. Although it would seem that they would minimize body damage (red-faced, we admit we crash tested one or two in the past), the hidden windshield wipers, GM's great Styling Schtick, introduce more damage potential in a front end collision by allowing the hood to jump back and crack your windshield — something that didn't happen with the old flush-with-the-deck-hoods. Also, in snow country the hidden wiper's well tends to pack up with snow and ice, rendering the wipers useless. Maybe this is one example why *stylists* per se are in the doghouse with some congressmen and consumer's groups.

The Esprit comes with a standard front anti-roll bar, which helps to minimize plowing around corners a bit but it looks as if you have to go up to the Formula model 'Bird to get the Handling Package, consisting of anti-roll

bars (1½ in. up front and ⅞ in. in rear), F60-15 treads, 15 in. x 7 in. wheels, high-rate rear springs and quick variable-ratio power steering. We're sure this great bundle of goodies fits the Esprit but, in an attempt to create some order back at the factory, Pontiac doesn't offer this kit for the Esprit.

In performance, the Esprit accounts for itself adequately with a best 0-60

Changes on the '72 Camaro consist of a toothier look up front and a sportier standard steering wheel. Z/28 is still hairiest model with SS 350 close behind. New trim trick is a "wet" vinyl roof in five colors.

time of 9.9 seconds, slower than the Javelin 401 four-barrel, but equal to the Camaro 350, which also has a two-barrel. Its best quarter-mile e.t. was a 17.3 at 81 mph, faster than the smaller-engined Camaro but slower than the Javelin by more than a second.

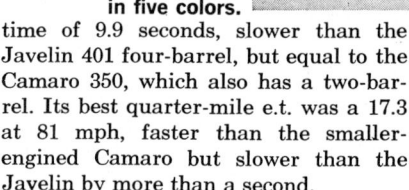

Firebird's Esprit is one step below the scooped-hood Formula model. Standard engine is 350 two-barrel with 400 two-barrel optional. Endura front bumper is standard. Four-on-floor is option. Standard front brakes are disc.

'CUDA

In perhaps a final gesture of Change For The Sake of Change, Chrysler restyled the taillights and grille of the Barracuda. The grille is now elongated horizontally instead of divided into vertical cubbyholes and Camaro-style round taillights replace the old ones.

The interior of the Barracuda, while very efficient in its character, struck more than one staffer as a little too

'72 Javelin comes with choice of four V8s, from 304 to 401 cubic inches. Only style change for '72 was new grille and optional rally stripes. Standard interior is vinyl, with corduroy fabric optional.

angular in its lines — foam-filled plastic seems to be jutting at you from all directions and even the seats appear to be harder than in the other ponycars. The "sharp edges" are, of course, an illusion because they are on crushable surfaces but they still don't *look* friendly. One item we wish they'd change is the inside door handles, recessed into the door grip. They have a tendency to pinch fingers — this in the third year of the car's development.

The 'Cuda 340 is still the same tough, lean car it always was, even if the 340 engine itself has been tamed considerably, emasculated by an 8.5 to 1 compression ratio, so that its horsepower rating has plummeted below the 275 gross horsepower it was rated at back when it had a 10.5 to 1 compression ratio.

Our Barracuda had a Hurst-assisted 4-speed, rather than the "Slap-Stick" shifter you can get with the Torque-Flite. It also had a 3.55 final drive, part of the performance axle package available on the 'Cuda. Other bits and pieces of the package are Chrysler's Sure-Grip differential, a seven-blade torque drive and radiator with a fan shroud.

The year 1972 will not be, as you have guessed by now, a bonus year for performance cars. Our test results confirmed it, with our Barracuda turning a 16.3 e.t. in the quarter, almost 2 seconds slower than the 340 Barracuda we tested way back in May 1970 but respectable considering how much it had to change to meet the regs. Of course, *that* old 'Cuda had a 10.5 to 1 compression ratio and a 4.10 to 1 rear axle ratio, plus wider (E60) treads than our '72 test car. But high (numerically) axle ratios are a thing of the past, too, at least if you try and order them on a new car (of course, you can order a replacement ring and pinion for a '70 or '71 'Cuda and install it on your '72) since they too affect emissions. What you see is what you get, etc. etc.

In cornering, the 'Cuda has what it takes to belong to the Ponycar Club, with anti-roll bars both fore and aft offered only on the 340. These minimize

lean in cornering and help to stave off that I'm-going-to-roll-over feeling. Unfortunately, higher cornering power is limited by the lack of treads wider than F70s but at least you know F60s will fit under the wheelwells since Plymouth has offered them in the past.

JAVELIN

The Javelin was our sole entry from

AMC, and, in our opinion, one of the most balanced of the ponycars. Just as it always has been. We ordered it with the 401-cu.-in. V8, which comes in a four-barrel form only, and was rated at 330 gross hp last year. We mated this to AMC's 3-speed Torque-Command Automatic, actually built by Chrysler. The rear end ratio was a tall 2.87 to 1 gear — great if you plan to do some long distance traveling but not exactly suited toward enhancing the car's supercar image around town. The tires on our test car were D78-14s, AMC's second widest tires, after C78-14, but far narrower than their only super-wide tread, the E60-15s standard on the AMX "Go" package.

The Javelin's styling provoked controversy among the staff, splintering them into "pro" and "con" groups. The "pro" group felt that the Jav has at last reached a maturation in styling with overall good lines highlighted by subtleties like the slight spoiler lip in the roofline or the front air dam on the AMX. The "con" group felt that the front fender bulges were an afterthought that looked too Kalifornia Kustom-ish and that the whole car offered nothing visually you couldn't match with almost any of the other supercars.

We know you won't believe this, but somehow the Javelin stomped all the other cars on the drag strip, although it beat the 'Cuda by only one-tenth of a second. We are at a loss to explain this, especially when you consider some of the other cars had a lower (higher numerically) rear end more suited for acceleration. Its being an automatic may have helped — more torque multiplication at low speed. Also, many of the other cars were two-barrels and, except

>>>>

SPECIFICATIONS

SPECIFICATIONS	CAMARO	JAVELIN	FIREBIRD	'CUDA	MUSTANG
Engine	90° OHV V8	90° OHV V8	90° OHV V8	90° OHV V8	90° OHV V8
Bore & stroke — ins.	4.00 x 3.48	4.165 x 3.680	4.120 x 3.746	4.40 x 3.31	4.00 x 3.00
Displacement — cu. in.	350	401	400	340	302
HP @ RPM	165 @ 4000	255 @ 4600	175 @ 4000	240 @ 4800	142 @ 4000
Torque: lbs.-ft. @ RPM	280 @ 2400	345 @ 3300	310 @ 2400	290 @ 3000	200 @ 2600
Compression Ratio/Fuel	8.5:1/Regular	8.5:1/Regular	8.0:1/Regular	8.5:1/Regular	9.0:1/Regular
Carburetion	2v	4v	2v	4v	2v
Transmission	3-spd auto, Turbo Hydro	3-spd auto	3-spd auto, Turbo Hydro	4-spd	3-spd SelectShift Cruise-O-Matic
Final Drive Ratio	2.73:1	2.87:1	3.42:1	3.55:1 with Sure-Grip	2.79:1
Steering type	Variable ratio power, ball-nut	Variable ratio power, ball-nut	Variable ratio power, ball-nut	Power	Variable ratio power, ball-nut
Steering Ratio	15.5-11.8:1	16.1-12.1:1	16.1-13.1:1	19:1	16.4-20.2:1
Turning Diameter (curb-to-curb-ft.)	39	36.3	36.5	N.A.	39.8
Wheel Turns (lock to lock)	2.29	3.2	3.3	3.5	3.17
Tire size	E78 x 14	D78 x 14	F78 x 14	F70 x 14	F70 x 14
Brakes	Power disc/drum	Power disc/drum	Power disc/drum	Power disc/drum	Power disc/drum
Front Suspension	Independent, coil spring, front stabilizer bar	Independent, unequal length upper and lower arms, coil springs and struts	Independent, coil spring	Independent, lateral, non-parallel control arms with torsion bars	Independent, coil spring
Rear Suspension	Multi-leaf spring	Multi-leaf spring	Multi-leaf spring	Asymmetrical semi-elliptical leaf spring	Multi-leaf spring
Body/Frame Construction	Integral body-frame, separate ladder-type front frame section	Unitized	Integral body-frame, separate ladder-type front frame section	Unitized	Unitized
Wheelbase — ins.	108	110	108	108	109
Overall length — ins.	188	191.8	191.6	186.6	189.5
Width — ins.	74.4	75.2	73.4	74.9	74.1
Height — ins.	49.1	50.9	50.4	50.9	50.8
Front Track — ins.	61.3	59.7	61.6	59.7	61.5
Rear Track — ins.	60.0	60.0	60.3	60.7	61.0
Curb Weight — lbs.	3,336	3,300	3,241	3,625	3,300
Fuel Capacity — gals.	18	16	18.6	19.0	20
Oil Capacity — qts.	4	4	5	4	4

PERFORMANCE

PERFORMANCE	CAMARO	JAVELIN	FIREBIRD	CUDA	MUSTANG
Acceleration					
0-30 mph	4.1	3.7	3.7	3.5	3.9
0-45 mph	6.7	5.6	6.5	5.5	6.7
0-60 mph	10.2	8.3	9.9	8.5	10.4
0-75 mph	18.5	11.5	15.1	12.8	16.2
Standing Start					
¼-mile Mph	79	89.5	77	85.5	78
Elapsed time	18.5	16.1	17.6	16.3	17.5
Passing speeds					
40-60 mph	6.2	4.2	5.7	3.9	5.7
50-70 mph	7.6	4.3	6.7	4.8	6.7
Stopping distances					
From 30 mph	23.8	33.5	34.3	25.7	33.5
From 60 mph	123.4	152.5	129.7	137.8	155.6
Speedometer error Electric speedometer Car speedometer	30 40 50 60 70 80 / 28 38 45 55 65 74	30 40 50 60 70 80 / 30 40 50 60 70 80	30 40 50 60 70 80 / 30 40 49 59 69 80	30 40 50 60 70 80 / 29 38 48 58 68 77	30 40 50 60 70 80 / 32 43 55 66 77 88

*Speeds in gears are at shift points (limited by the length of track) and do not represent maximum speeds.

ROUNDUP

for some guys in Arizona who race a two-barrel GTO on the strip, two-barrels are not the hot setup for performance, at least when you only put one on an engine . . .

In braking, the Javelin's power discs hauled the car down from speed well enough but, just like the others, in a 60-mph panic stop, there was some wheel hop in back and some directional correction required by the driver to keep the car in one lane. This brings up a sore point about the ponycars — why were they never really refined with features like four-wheel discs and maybe even independent rear suspension? "Cost," is too obvious an answer, for a Camaro or Javelin with both those features could justifiably go for $1,000 more. Perhaps that's another reason the ponycars are dying — because their looks promised more than they could deliver. Even a Subaru or Fiat can take a corner as well as the average "economy" ponycar and maybe the enthusiasts got wise and bought cars that

The original ponycar, the Mustang, offers either a "flatback" or "tunnelback" roofline for '72. Top engine is 351-cu.-in. HO "Boss" engine. Other V8s are two 351s and a 302. A four-speed and ram induction are optional.

performed first and looked good second.

Power plants for the Javelin permit it to span a wide range of customers. There are two sixes, a 232 and a 258-inch, a 304-cu.-in. two-barrel V8, two 360-cu.-in. V8s, one with a two-barrel carb and one with a four, and finally the 401-cu.-in. V8, the one in our test car. Unfortunately, none of the AMC engines, other than the 360 four-barrel, have gained much of a reputation among performance enthusiasts, most of whom think of engines like the Z/28, Boss 302 and 351 or Mopar 340 when they think of hot small blocks.

MUSTANG

From Ford, we obtained a '72 Mustang hardtop, which, in keeping with the Mustang's gradual spreading at the hips, arrived with plushy "Grande" trim, which includes a vinyl roof, cloth seats, simulated wood and various sparkle-plenty moldings.

The power plant was Ford's 302-cu.-in. V8, the same engine that's standard in the Mach I and the first option for all the other Mustangs (not counting the hairy-chested Boss 351). This is the same engine which last year had a 220

SAE gross horsepower rating at 4600 rpm with 300-lbs.-ft. of torque at 2600 rpm. It runs a two-barrel carb with an automatic choke and burns regular gas. Back in '70 the 302 had a 9.5 to 1 compression ratio. For '71, it was lowered to 9.0 to 1, for even better compatibility with non-leaded gas. In time, there won't be a premium gas pump in the land . . .

The '72 Mustang strikes anyone who remembers the first Mustangs kind of funny, almost as if it were a fattened up caricature of the first 'Stangs. Everything's bulgier, as if Miss America 1964 suddenly reappeared, carrying 40 more lbs. of ugly fat. Even the interior seems to have huge expanses of plastic stretching across your horizon to no particular purpose. But there is a functionality of sorts, principally in the grouping of all heater, air conditioner, radio and miscellaneous controls in a vertical console between the two front bucket seats.

Our nit-picks on the Mustang interior? The central console, sort of a combination armrest/reserve glove compartment, was handy as an armrest but it seems only slightly taller than The Great Wall of China. Second, the Mustang's inside rear-view mirror manages to block out the upper third of a tall driver's visibility through the center of the windshield. This is enough to totally lose sight of a pedestrian when turning a corner. Our last bone to pick with Ford's product planners concerns the air conditioning. It's not that it dragged the engine power down noticeably, as happens on the real minis like the Pinto. It's just that the center console vents don't turn from side to side, a bit frustrating when you want the cool air pointing at you rather than empty space. Better they should go to the revolving sphere that GM puts in the ends of their vents.

The '72 Mustang's handling was surprisingly good, keeping up that tradition of cat-quick steering that all Mustangs have had since the beginning, even the sixes. Of course, ours had

power steering, which made things even easier, but we were really pleasantly surprised that a Mustang without the 351's Competition Suspension could still go right where you pointed it. Fortunately, you can order the Competition Suspension for any of the Mustang V8s and we would, since you never know when you might have to turn a corner rather suddenly and would prefer to do it staying right side up.

In drag-strip performance, the '72 Mustang 302 with its three-speed automatic proved to be a bit slower than the '71 Mustang 302 we tested in January. Our '72, for instance, went from 0-60 mph in 10.3 seconds while back in '71, we did it in 9.9 seconds. The quarter-mile e.t. was the same, 17.5 seconds, while the speed in the quarter was slightly slower than last year, only 75 mph, by the time it tripped the lights.

Now comes the time for judgment. We gather *Motor Trend* readers, who now have been through many multi-car tests with us, want a definite answer at the end — a pinpointing of which car we would recommend *over all*. Backed thusly against the wall, the majority of the staff said they would choose the Mustang over the other four we tested. Why? Admittedly, it wasn't the fastest. And its styling is not as slick as some of the others. Its cornering was average, too. But, perhaps reflecting the view of today's ponycar buyer, we asked ourselves, "What do we get for our money?" The sad truth is, although we were mixing apples and oranges in our road test by having a big-engined Jav pitted against a small-engined Camaro and so forth, we knew *evenly matched versions* of the "hottest" '72 ponycars couldn't hold a candle to the '70s. A '70 Z/28, AAR 'Cuda, Boss 302, etc. would blow their doors off. So what's left? The ponycar as an economy car. There is where you get your money's worth for '72. The wide range in options, engine sizes and prices of our '72 ponycars helped convince us you're not really getting much more if you spend more dollars in '72. We'd venture to say, with any of the brands tested, an economy version similar to the Mustang's setup would be the wisest investment. It's almost ironic, with the larger engines and flashy trim to choose from, most of the staff seemed content with the car which came closest to the original marketing concept developed for Mustang way back in '64 — "a compact, economical car with a sporty flair." /MT

New Barracuda has choice of V8s limited to 318 or 340, with new electronic ignition. Scoops are standard on 'Cuda performance model. Four-speed was best we've shifted on ponycars to date.

PONYCAR ANNUAL SALES FIGURES

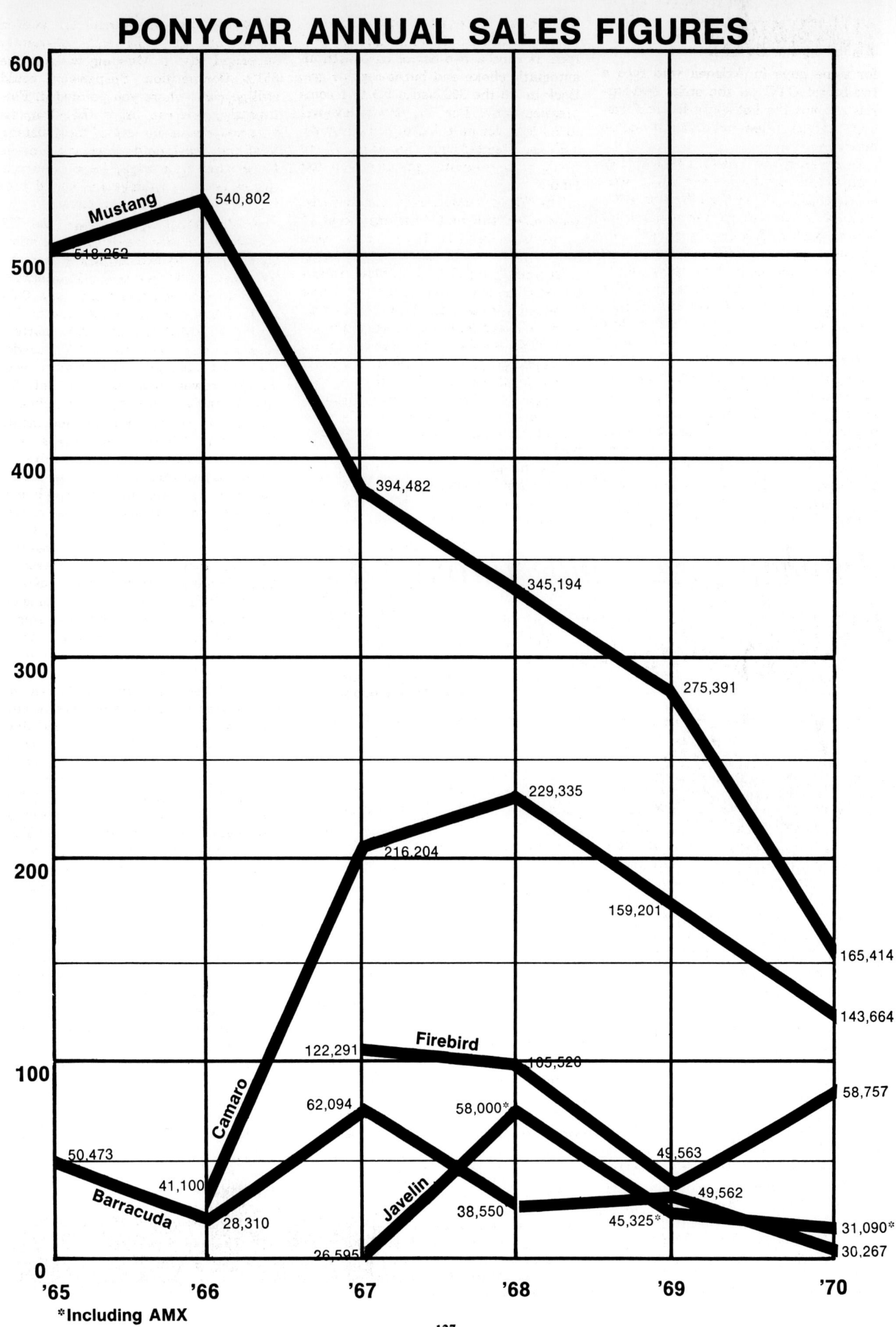

600

Mustang

518,252

540,802

500

394,482

400

345,194

300

275,391

229,335

216,204

200

159,201

165,414

143,664

Firebird

122,291

105,520

Camaro

100

62,094

58,000*

58,757

50,473

49,563

41,100

49,562

Barracuda

Javelin

38,550

45,325*

31,090*

28,310

30,267

26,595

0

'65　　　　'66　　　　'67　　　　'68　　　　'69　　　　'70

*Including AMX

127

CHUCK BOONE PHOTO

CAMAROS FOR EVERYTHING

An engineered guide through a thicket of options for luxury, performance or combinations thereof

THE BEWILDERING ARRAY of options with which the domestic factories bemused and bedazzled the public for so many years is withering away, victim in equal measure of public indifference and government controls. There are fewer models that differ in name only, fewer engines that vary only in rated horsepower, fewer inconsequential variations that nobody cared about anyway.

A good thing all around, for while the indifferent public is less subject to confusion, the more enlightened factories have a new opportunity to do what they call "engineering packaging," meaning that the men responsible for designing the various engines, transmissions, suspensions, interiors, instrument panels, trim and so forth now have more influence in deciding what should go with what and (more important to the buyer, who usually takes what's on the showroom floor) the designers can help the dealers order the proper combinations. Less chance for the customer to get a bad combination, which he could easily do in the days of design-it-himself.

Our evidence of this is in the form of three Chevrolet Camaros. Chevrolet is not the only factory involved with engineering packaging. All the domestic manufacturers offer trailer-towing or performance combinations for most models. But Chevrolet seems to have gone to the most effort and has applied that effort to a model we like very much.

The three examples were alike in dimensions and all of course were coupes, the only body style in which Camaros are made. Each had the Rally Sport exterior, mostly a matter of two separate (and more attractive) front bumpers instead of the solid bar dividing the eggcrate grille.

Our engine selection consisted of only two versions of 350 cu in. V-8; the 2-barrel 350 with 165 bhp (in one car with automatic transmission and in a second car with 4-speed manual) and the 255-bhp engine, gained with a higher compression ratio, wilder camshaft, 4-barrel carburetor and me-

chanical lifters (in a Z28). The choice was narrower than we liked because several Camaro engines, i.e. the 307-cu-in. V-8 and 400-cu-in. V-8, aren't certified for sale in California. The 350 V-8 with hydraulic lifters, milder cam and 4-barrel was offered, but we've tested it before and it would have obscured the direct comparison between automatic and 4-speed. The reader may assume, based on the earlier test, that the SS350 (so named because it comes with the SS option package) will knock about one second from the 1/4-mile time in comparison to the 2-barrel 350 and cost practically nothing in fuel economy in everyday driving.

What we got in our three package engineered Camaros, then, and the names assigned to each by the engineers who conceived the combinations, were the following cars:

The Budget GT had the 350 cu in. 2-barrel V-8. There were two choices of transmission and final drive ratio: wide-ratio 4-speed and 3.08:1 gearing or 3-speed automatic with 2.73:1 final drive. We picked the former for this car because we wished to try the newly-revised gearshift linkage and liked the idea of four widely-spaced ratios and highway gearing. For the chassis there was the sports suspension (F41 on the order blank) consisting of a larger front anti-roll bar, a rear anti-roll bar and stiffer rear shock absorbers, 14 x 7 styled steel wheels with F70 bias-belted tires, an instrument panel with a full set of gauges, power brakes and steering, a limited-slip differential and an AM/FM radio. The intent of this package, as the name suggests, is to provide extra performance without raising the price beyond what a young enthusiast could afford either in first cost or insurance.

The Luxury GT had the same engine but used Chevrolet's normal 3-speed automatic transmission (M40) and the standard final drive for this transmission, 2.73:1. The car had power brakes, power steering, rally wheels and F70 tires but standard suspension. This one too lived up to its name with

PHOTOS BY GEOFFREY GODDARD

CAMAROS FOR EVERYTHING

air conditioning, electric clock, rear window defroster, and custom interior (Z87). This last comes in two parts; acoustical improvement from additional floor, tunnel and roof insulation, a noise dam between fenders and cowl and a fiberglass blanket glued to the underside of the hood, plus extra trim and cloth instead of vinyl on the seats. An interior lighting group is part of the option. There was a tiltable steering wheel and the standard, rather vacant, instrument panel.

The Performance GT was the old familiar Z28, somewhat the worse for emission controls. The Z28 option is the most extensive in the line, and includes the 255-bhp engine with a choice of close- or wide-ratio 4-speed or a 3-speed automatic with torque converter and shift points revised to suit the more powerful engine. We picked the specially tuned automatic for contrast and used the 4.10:1 gearing rather than the also available 3.73 in order to obtain the best possible acceleration. Also part of the Z28 option are a limited-slip differential, a larger radiator, a flexible cooling fan, and 15 x 7-in. wheels with 60 series bias-belted tires. The package includes larger anti-roll bars than come with the normal handling option, and it has shocks and spring rates tuned to work with the larger tires. On the test car and in keeping with the theme were front and rear spoilers (D80) and a few concessions like tiltable steering wheel and custom interior. This is as far as Chevrolet is willing (or able?) to go with performance.

The test data clearly show that Chevrolet's primary aim was squarely hit. The Budget GT is the least expensive, strikes a balance in acceleration between the luxury and performance models and gets the best fuel economy, as it should. The Z28 was the quickest, had the highest cornering power and was the most visible. And the luxury car was the slowest, the most expensive and the quietest.

The subjective results are nearly the same. The 2-barrel 350 V-8 and 4-speed turned out to be a most satisfactory combination, quiet and at least acceptably smooth for a 1972 emission-controlled engine. There was some evidence of the carburetion leanness to which the engine has been subjected, especially at low speeds and light throttle. There was a tendency to stall when cold, but few 1972 engines have not been affected this way. With the numerically low gearing it didn't pull strongly in top gear but the choice of intermediate ratios made up for that. Close ratios are not the best choice for daily driving, despite folklore. And the shift linkage is a pleasure to operate, in contrast to the overly stiff Camaro linkage of earlier years.

The variable-ratio power steering is quick, albeit lacking in feel. The F41 suspension stiffens the ride but it is responsive and provides a useful increase in cornering power. The extra instruments (tachometer, water temperature gauge, ammeter and electric clock) are appreciated. (Forgot to say that's option U14.) And the steering wheel (NK4), small and with a vinyl-covered rim, adds to the pleasure by providing a good driving position and grip.

The Z28 is perhaps less successful. As mentioned, it is the quickest of the three and the quickest passenger car (that is, excluding the Corvette) that Chevrolet sells. However, this no longer means what it once did. All the domestic manufacturers have withdrawn from the horsepower race so the 1972 competition isn't much. Thing is, the 1972 Camaro is no match for the 1970 Camaro. The present horsepower rating is accurate while the earlier ones weren't, so don't assume the Z28 has actually dropped from 360 bhp to 255, but even so, the 1970 engine, with 11:1 c.r. and less stringent emission regulations to meet, had a net output of about 300 bhp, or 45 more than the 1972 engine. And it shows. The

1972 CAMARO SPECIFICATIONS

Engine

Type	ohv V-8
Bore x stroke, mm	101.6 x 88.4
Equivalent, in	4.00 x 3.48
Displacement, cc/cu in	5735/350

Chassis & Body

Layout.............................front engine, rear drive
Body/frame.............unit steel with front subframe
Brake type...............:11.0-in. vented disc front, 9.5 x 2-in. drum rear; vacuum assisted
Swept area, sq in.....................332
Steering type: recirculating ball, power assisted
Overall ratio...................15.5/11.8:1
Turns, lock to lock.....................2.3
Turning circle, ft......................41.1
Front suspension: unequal-length A-arms, coil springs, tube shocks, anti-roll bar
Rear suspension: live axle on leaf springs, tube shocks, anti-roll bar (optional)

Accommodation

Seating capacity, persons...............2+2
Seat width front/rear........2 x 22.5/2 x 20.0
Head room, front/rear.............37.0/35.0
Seat back adjustment, degrees..............4

General

Wheelbase, in.........................108.0
Track, front/rear...................61.3/60.0
Overall length........................188.0
Width...............................74.4
Height..............................50.5
Ground Clearance......................4.5
Overhang..........................38.1/41.9
Usable trunk space, cu ft..................6.5
Fuel tank capacity, U.S. gal.............18.0

ROAD TEST RESULTS

Acceleration	Luxury GT	Budget GT	Z28
Time to distance, sec			
0–1320 ft (¼-mi)	17.6	17.2	15.5
Speed at end of ¼-mi, mph	79	82.5	90
Time to speed, sec			
0–30	3.9	3.5	3.3
0–60	10.5	9.8	7.5
0–90	25.0	22.0	15.5

Speeds In Gears			
Top gear	107	110	124
3rd		85	
2nd	92	69	84
1st	56	49	48

Brakes			
Minimum stopping distances, ft.			
From 60 mph	161	161	159
From 80 mph	300	300	295
Pedal effort for 0.5g stop, lb	25	25	25
Fade: percent increase in pedal effort to maintain 0.5g deceleration in 6 stops from 60 mph	100	100	100
Overall brake rating	good	good	fair

Handling			
Speed on 100-ft radius, mph	32.4	32.9	33.2
Lateral acceleration, g	0.702	0.723	0.736

Fuel Economy			
Normal driving, mpg	13.6	15.1	12.3
Cruising range, mi	236	257	209

Interior Noise			
All noise readings in dBA:			
Idle in neutral	55	56	62
Constant 30 mph	65	65	67
50 mph	68	71	73
70 mph	74	74	76
90 mph	77	78	81

earlier car, with the same gearing and transmission, covered the standing quarter mile 1.5 sec quicker than the present car.

That problem aside, there are some other problems. The idle speed is set fast and the transmission usually goes into gear with a bump. Even at the high speed the idle is rough, and the special pistons and mechanical lifters set up noises of their own. These low-compression, high heat-rejection engines may *run* on 91-octane fuel but they don't always stop on it and the Z28 was liable to run-on when warm.

Let's modify some of this. The Z28 isn't all bad. The engine is fitted with an air pump, a benefit because the pump allows fewer compromises elsewhere, and the highest-performance engine is less bothered by leanness and surge than the milder 350 in our other two cars. And the car does move briskly, aided by the good low-speed torque and revised torque converter. It fairly leaps from rest and gains speed at a good rate, shifting at redline with a nice touch of wheelspin. The suspension and large tires provide high cornering power and the Z28 can be driven fast with precision.

If the 4-speed manual transmission on the Budget GT was an unexpected bonus, then the automatic transmission in the luxury car is no detraction at all. That is, you gain something with the stick yet lose nothing by not having the stick. The abundance of low-speed torque moves the car from rest as quickly as the driver could wish, without the strained feeling typical of a European sporting sedan with a relatively small engine tuned for high output. The Turbo Hydra-Matic supplied with the 2-barrel 350 has a lower stall speed, lower shifting points and softer shifts than the

one in the Z28, in keeping with the luxury car's character. The engine and transmission combine to give effortless, noiseless and turbine-smooth performance. The unexpected bonus here is that the automatic transmission dampens the surge caused by the required lean mixture. A stumble on acceleration from idle when cold was the only readily noticeable symptom of the 1972 controls on this car.

The standard suspension is softer and the luxury car had lower cornering power and more initial understeer, both predictable. It also *jiggled* less on rough surfaces. The optional interior is attractive without being gaudy; we missed the full set of instruments although probably the luxury-oriented buyer will feel more secure with warning lights. The air conditioning (not available with the Z28) gets an unequivocated endorsement. This is a field in which Chevrolet (and GM) is indisputably skilled. There is no better unit on the market unless one considers the automatic temperature control units on larger U.S. cars essential.

Not all the options are worthwhile. The Z28 was not, as you may have noticed, our favorite. There wasn't enough blinding speed to make up for the fuel consumption and engine temperament. All three cars had the optional adjustable seatbacks, with a range of only 4°, and the backs are too upright to begin with. It's a good idea, insufficiently executed. And we resent having to buy the interior brightwork and applique as a condition for getting the extra sound insulation. The latter is worth the money, the former less so.

And there are options Chevrolet should offer but doesn't. At one time the big Caprice could be bought with a steering-column stalk control for the windshield wipers. Caprice

ENGINE AND DRIVETRAIN

Engine	Luxury GT	Budget GT	Z28
Compression ratio	8.5:1	8.5:1	9.0:1
Bhp @ rpm	165 @ 4000	165 @ 4000	255 @ 5600
Equivalent mph	108	99	108
Torque @ rpm	280 @ 2400	280 @ 2400	280 @ 4000
Equivalent mph	65	59	77
Carburetion	Rochester 2V	Rochester 2V	Rochester 4V
Fuel required	reg, 94-octane	reg, 94-octane	reg, 94-octane
Drivetrain			
Transmission type	3-speed auto	4-speed manual	3-speed auto
Final drive ratio	2.73:1	3.08:1	4.10:1
Wheels	14 x 7	14 x 7	15 x 7
Tires	F70-14	F70-14	F60-15
Mph/1000 rpm (top gear)	27.1	24.6	19.2
Engine revs/mile (60 mph)	2220	2430	3120

CAMAROS FOR EVERYTHING

buyers didn't care, which we could have told Chevrolet beforehand. Camaro buyers would appreciate it and we wish Chevrolet would offer it to them. (When this happens we will begin agitating for a headlight dimmer stalk control. Never satisfied.)

Let us now turn these specialized models around. Is there a minus for each plus? Not really. The Budget GT is not a plain-pipe-racks sort of car. For the improved handling and driving pleasure you lose little in the way of ride comfort

and quietness. The Z28 is also quiet as sporty cars go, though the sound level meter reveals it to be significantly noisier than the other two. The suspension and big tires harshen the ride but not too much. And the luxury GT has better than acceptable handling and performance. All have good brakes, though fade resistance is marginal for the Z28's acceleration capability. It's all a matter of degree, or perhaps balance. The budget car doesn't lack luxury, the performance car doesn't lack comfort and the luxury car doesn't lack performance.

We didn't pick a favorite from this group. Our choice of Camaro would be the SS350; with 4-barrel carburetor and hydraulic lifters it's quieter than the Z28 yet more powerful than the 2-barrel version. We'd also get the F41 suspension. Either transmission would be fine and we wouldn't want to leave out that good air conditioning.

That's us, and that's engineering packaging. It's good to know the options are available, better to have the factory helping put the right combinations in dealer showrooms and best that the drawbacks seem less than the advantages, whichever way you go.

PRICES

Budget GT

List Price, $2776; as tested $3850.00
Price as tested includes sport suspension, $30.55; positraction axle, $45; 165-bhp engine, $26.35; power brakes, $47.40; 4-speed transmission, $200; power steering, $113; F70 x 14 tires, $68.50; special instrumentation, $84.30; AM/FM radio, $135; 14 x 7 wheels, $44; Rally Sport package, $118.

Luxury GT

List price, $2776; as tested $4364.85
Price as tested includes tinted glass, $39; rear window defroster, $31.60; air conditioning, $397; power brakes, $46; 165-bhp engine, $26.35; automatic transmission, $210; tiltable steering wheel, $44; power steering, $113; AM/FM radio, $135; F70 x 14 raised letter tires, $83.60; Rally Sport package, $118; custom interior, $113.

Z28

List price, $2776; as tested $4558.45
Price as tested includes Z28 option (255-bhp engine, dual exhausts, handling package, 15 x 7 wheels and F60 x 15 tires, stripes and emblems, $766; automatic transmission (required with Z28 option but priced separately), $297; center console, $57; front and rear spoilers, $77; power steering, $113; AM radio, $65; Rally Sport package, $121.15; custom interior, $113; 4.10:1 axle, $12.

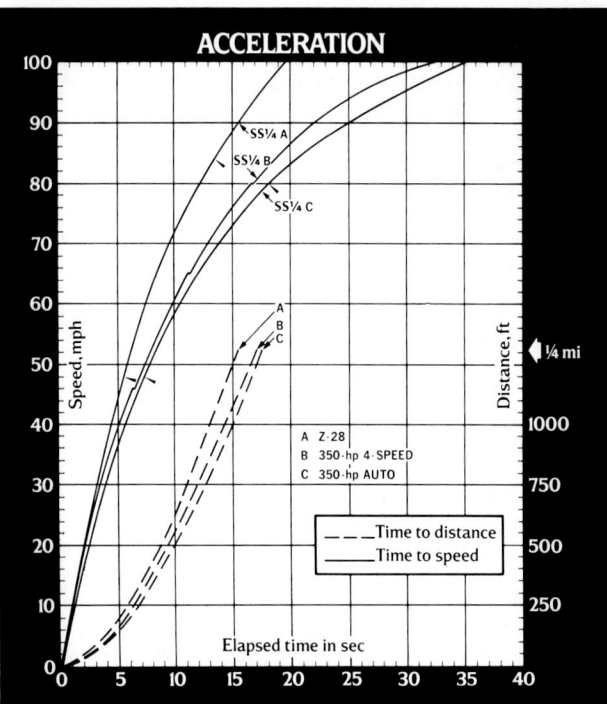

ACCELERATION

A Z-28
B 350-hp 4-SPEED
C 350-hp AUTO

- - - Time to distance
——— Time to speed

SLEEPER SIX

By C.J. Baker ■ If you're cruising the streets of Detroit and you pull up beside a '72 Camaro with trick gold paint and small script proclaiming "Turbo Camaro," beware! It's a six-banger. "No sweat," you say. "Six in row just won't go." But don't be fooled, this six runs like a 350-cubic-inch V8 with a four-barrel. It's Chevrolet Engineering's latest experiment with turbocharging, and while there are no plans to market such a vehicle, this Camaro and a twin-turbocharged Nova, reviewed in the March issue of our sidekick publication, *Car Craft,* are being used to evaluate the virtues and drawbacks of turbocharging in relation to performance emissions, reliability and economy.

The real beauty of the six-cylinder adaptation is that it is a bolt-on installation, with the single exception of brazing a fitting into the side of the oil pan for the oil return line from the turbocharger. The only modification that Chevy made to the engine was the substitution of stock export truck pistons, which have a compression ratio of only 7.0-to-1 as compared to the production 8.5-to-1 pistons. Additional performance could have been achieved if the

A six that runs like an eight, good emissions, reliability and economy too? That's Chevy's new experimental Turbo Camaro

8.5-to-1 pistons had been retained, but premium fuel would have been required to eliminate the possibility of detonation. But since the current GM policy is to configure all engines to operate on no-lead fuel, the 7.0-to-1 pistons were installed.

The turbocharger, a Schwitzer Model No. LDA319, is bolted to the intake manifold, using a sandwich-like clamping bracket. This particular turbocharger, utilizing a 3-inch-diameter turbine, is capable of supplying 319 cubic feet of air per minute at a 2-to-1 pressure ratio while operating at speeds up to 80,000 rpm. Intake air for the compressor section of the turbocharger is supplied through a 3-inch-diameter flexible rubber hose which opens into the grille. From the compressor, the pressurized air, supplied at

up to 10 psi of boost pressure, is routed to a Vega air cleaner atop a Rochester Monojet single-barrel carburetor. The air enters the Vega air cleaner by way of an elbow borrowed from a turbocharger kit produced by Schwitzer for the Vega. This elbow is also equipped with a fitting for attaching a boost pressure sensing hose which controls a Holley electric fuel pump. The Holley pump, which maintains fuel pressure at 4 psi above the air pressure in the air cleaner, is used only when additional fuel pressure is needed, supplementing the stock mechanical pump which runs continuously. This pumping arrangement assures that fuel will flow properly into the carburetor float bowl and through the carburetor regardless of the internal pressure inside the carb due to the turbocharger.

The Rochester Monojet carburetor is basically stock, although the power circuit has been enriched slightly. The idle circuit is unaltered, and the engine retains the standard A.I.R. pump to keep the emission levels within bounds. It was also necessary to reroute the crankcase breather hose from its normal attachment to the air cleaner to the intake side of the compressor,

ABOVE LEFT — A completely bolt-on package, the turbocharger fits neatly into the Camaro engine compartment. ABOVE — A Vega air cleaner with a special elbow and a pressure sensing fitting to control the electric fuel pump is used. BELOW LEFT — Holley fuel pumps maintain fuel pressure at 4 psi above boost pressure. RIGHT — Photos show the flexible steel line which carries oil from a tee installed at the oil pressure sender switch to the turbocharger. Oil gravity feeds back to the engine via a separate line. BELOW — Gold paint and white accent stripes highlight Turbo Camaro. Wheels are stock Z/28 units painted white to match the car.

photography:
Mike Brenner

where it is continually exposed to a negative pressure area to ensure proper crankcase ventilation.

To drive the turbocharger, a short U-shaped exhaust pipe is used to route the exhaust gases from the exhaust manifold to the turbine. After leaving the turbocharger, the exhaust passes through another exhaust pipe which connects to the existing stock exhaust system about halfway back under the car. The Turbo Camaro was fitted with a Z/28 dual-inlet, dual-outlet transverse muffler, utilizing only one inlet, but both outlets. Thus the car appears to have dual exhausts. This modification was made for appearance reasons and for a more pleasing exhaust sound.

Since the turbocharger operates at such high speeds, adequate lubrication is essential. To handle the lubrication chores, a flexible steel oil line is run from a tee fitting installed between the oil pressure warning light sender switch and the engine block. Hence oil is supplied to the turbocharger at full engine oil pressure. After leaving the turbocharger, the oil gravity feeds back into the engine oil pan through a separate line.

To get a real appreciation of just what the turbocharger was doing for the Chevy six, we checked out a stock six-cylinder Camaro to take for a drive before testing the Turbo Camaro.

The only difference between this car and the turbocharged unit, other than the turbocharger, was that the stocker had a powerglide transmission whereas the Turbo Camaro was equipped with a turbo-hydro. In the stock form, the Camaro thundered through the quarter-mile in a heart-stopping 20.28 seconds at 67 mph — just about time enough to go get a hot dog and a cold drink before the next round.

Next, we climbed into the Turbo Camaro. The turbocharged engine started, idled and ran every bit as smoothly and quietly as the stocker, but that's where the similarity ended. When you pushed down on the accelerator, you got more than just exercise for your ankle. Engine response was quick and smooth without hesitation. Quarter-mile runs produced a best of 16.18 seconds at 86 mph — not exactly a AA/Fueler, but still more than 4 seconds and nearly 20 mph better than the

stocker. However, the real eye-opener was in the 50-to-70-mph acceleration times; the Turbo Camaro required only 5.12 seconds, as compared to 12.87 for the unturbocharged car. This kind of response is an unquestionable safety factor on today's highways, cutting the time required to pass a slower vehicle by more than half.

All in all, the Turbo Camaro was a pleasant car to drive, exhibiting none of the ill manners that are usually associated with modified engines. Considering the simplicity of turbocharging, the improved exhaust emission levels, the performance and safety gains, we certainly hope that the Turbo Camaro, and its sister vehicle, the Twin-Turbo Nova, will generate more interest and experimentation with turbocharging for possible future production. If so, perhaps cars like this will find their way into your dealer's showroom and onto the streets of Yourtown, U.S.A. ■■

USED CAR TEST

No. 370

1970 Chevrolet Camaro Z28

PRICES

Car for sale at Kingston, Surrey, at	£2,390
Typical trade/cash value for same age and model in average condition	£1,950
Total cost of car when new including tax	£3,219
Depreciation over 3 years	£1,369
Annual depreciation as proportion of cost new	15 per cent

DATA

Date first registered	13 December 1970
Number of owners	1
Tax expires	31 March 1973
MoT	Not yet needed
Fuel Consumption	12 mpg
Oil Consumption	Negligible
Mileometer reading	24,800

PERFORMANCE CHECK

(Figures in brackets are those of the Chevrolet Camaro SS Used Car Test, published 12 August, 1971)

0 to **30** mph	**2.7** sec	(3.8)
0 to **40** mph	**3.6** sec	(5.5)
0 to **50** mph	**4.7** sec	(6.7)
0 to **60** mph	**6.3** sec	(8.5)
0 to **70** mph	**8.6** sec	(10.8)
0 to **80** mph	**10.8** sec	(13.8)
0 to **90** mph	**13.4** sec	(17.3)
0 to **100** mph	**16.5** sec	(22.0)
0 to **110** mph	**21.2** sec	(—)
0 to **120** mph	**26.6** sec	(—)
Standing ¼ mile	**13.6** sec	(16.3)

In top gear:

20 to **40** mph	**5.4** sec	(—)
30 to **50** mph	**4.9** sec	(—)
40 to **60** mph	**4.6** sec	(4.0)
50 to **70** mph	**4.8** sec	(4.0)
60 to **80** mph	**4.7** sec	(4.1)
70 to **90** mph	**4.7** sec	(5.3)
80 to **100** mph	**5.3** sec	(6.5)
90 to **110** mph	**6.8** sec	(8.2)
100 to **120** mph	**8.8** sec	(—)
Standing Km	**27.3** sec	(28.7)

TYRES

Size: F60-15in. Goodyear Polyglass GT

Approx. cost per replacement cover £18.70

Depth of original tread 7mm; remaining tread depth, 5mm on all tyres, including spare.

Tools

Bumper jack and wheelbrace. No handbook in car.

CAR FOR SALE AT:

American Car Centre, 144 London Road, Kingston-on-Thames, Surrey. Tel: 01-940 1733.

IN the 1960s, most American cars seen in Britain were vast, unwieldly lumps, totally unsuited for our narrow, twisting roads. Then Ford produced the Mustang, a compact, neat four-seater, with a sporting image and something approaching the handling we had come to accept. Chevrolet's answer to this was their version of the "pony" car, the Camaro.

This particular Z28 was imported from Canada only a few weeks ago, hence the L-suffix on the registration plate. In fact, the car is a 1970 model, with just one previous owner.

The 5.7-litre V8 engine, with solid pushrods, rather than hydraulic lifters, has been mildly tuned, to the extent of fabricated tubular exhaust manifolds ("Headers" in the States) and a rather better inlet manifold for the two-stage, four-barrel Holley carburettor to feed into. No claims are made for the power output of this particular engine, but from the performance figures it can be seen that it is quite adequate! Starting was always good, with one prod on the accelerator pedal setting the automatic choke. The noise is impressive, with a deep, powerful throbbing and a typical "sporting" V8 waffle from the exhausts. At the end of the test, two plug leads came adrift, but apart from a reluctance to tick over, there seemed to be little affect on the power output.

This is one of the few American cars we have had through our hands with a manual gearbox. This GM box, with a massive Hurst shifter, has a rather heavy action, but it is, at the same time, very precise. The gate protecting reverse, which is alongside first gear, has become rather weak, and care has to be taken to check that the right gear has been selected. As a safety precaution the starter is coupled to a cut-out on the clutch pedal, which prevents it from operating unless the pedal is depressed fully. The clutch movement is rather heavy, and the return spring squeaks.

By comparison, the disc front, drum rear brakes are ultra light at town speeds, needing just a touch to stop. At high speeds, they are superb, pulling the car down from 120 mph to a standstill without any roughness, weaving or signs of fade. A foot-operated parking brake is fitted, but with the seat belts correctly fastened, it is impossible to reach the release handle.

This Camaro is fitted with power steering, and with the combination of Goodyear Polyglass GT bias-belted tyres, allows the car to wander too easily. Despite an adequate amount of tread, the Goodyears have very little grip in the wet. They also tend to "white line" too much, so that the car needs steering every inch of the way. The dealers say that the best British tyres to fit are Dunlop F70 ER70-15 or Avon GR70VR15 radials.

The handling is good, remaining virtually neutral right up to the limit. There is very little roll, and while the suspension is on the firm side we felt that the car was a little under-damped.

With used cars, we do not drive them to the limit during performance testing. Too many things have a tendency to expire under this extreme treatment. But the Camaro did not need kid-glove treatment. On a dry track, the clutch was dropped at 4000 rpm and with a screech from the tyres, the car took off like a missile, nose up all the way to 120 mph, in a mean time of 26.6sec. The clutch is perfectly adequate for this sort of occasional treatment.

With the rather low gearing, the fuel consumption tends to be on the low side. In town and commuting, the figure dropped to fractionally under 10 mpg. In its more natural environment, cruising at 70 mph, this figure improved to around 12 mpg. As with most American V8s, virtually no oil was used.

In its present state, the car has combined

amber front indicators and side lamps, and red rear indicators. These can be quite easily converted to current 1973 Construction and Use Regulations.

The facia layout is reasonably clear, and the instrument lighting good. It is curious to see that despite the efforts of Ralph Nader, the heater controls are totally hidden from view by the steering wheel spoke. Ventilation and heating are good, but the car desperately needs a heater panel on the rear window to prevent misting.

This Camaro Z28 is an All-American car — yet compact enough to be used on Britain's roads, despite the shortcomings of left-hand drive. Its condition is excellent, and with a set of decent European radial ply tyres fitted, one would be much more confident of its wet-road behaviour.

Condition Summary

Bodywork

Were it not for one or two minor marks, it would be difficult to distinguish this car from a brand new one. The light moss green metallic paintwork has a deep lustre, set off by two wide black stripes running fore and aft. There are no visible signs of any major accident repair work. The brightwork is also in excellent condition, with no signs of marking or corrosion anywhere. Underneath, the condition is also good, with all the paint unmarked. The special exhaust system is free from rust, but one mounting bracket will need attention before long.

Like the exterior, the inside of the Z28 is "as new", with fitted rubber mats protecting the black carpet. The black seats are unmarked, and show little signs of use or wear. The knob on the window winder on the driving side has worked loose, and the window itself on the front passenger side needs to be wound up with the door slightly open to get the optimum seal.

Equipment

The speedometer is only *just* legal, under-reading by about 10 per cent at 30, improving to 5 per cent at above 60 mph. The headlamps have been altered for driving on the left, but the left-hand one is aimed too high. The only equipment not working are the twin reversing lamps and this is probably only a matter of replacing bulbs.

Accessories

A Delco medium waveband radio is fitted, wired to work through the ignition only. This is one of the few cars we have had with the aerial built into the windscreen. A poor connection, possibly with the radio earthing, caused a lot of interference unless the signal strength was high. A door mirror is fitted — and it would be useful if the twin could be put on the passenger door to cover the right-hand side of the car. Separate lap and diagonal belts are standard equipment, with the lap section having a self-stowing reel on its outboard end: this should not be mistaken for an inertia reel. If the ignition key is left in the car, an infuriating buzzer warns the driver when the door is opened.

About the Camaro Z28

It was in 1968 that General Motors in the USA decided to enter the TransAm series. In order to get the essential bits and pieces accepted, 1000 "homologation specials" were built. The original cars had 5.3-litre V8 engines, which would turn over at 7500 rpm. The 1970 cars had the 5.7-litre engine, with a rather lower rev limit. To make up for the power losses incurred by all the anti-pollution equipment, current Camaro Z28s have 6.6-litre engines, and in Britain are usually sold with automatic transmission and power steering. □

The Last Round-Up For Ponycars?

Not yet. The reports of the demise of the
sporty car are exaggerated . . . or at least a
bit premature/By Jim Brokaw

The Ponycar market is as difficult to understand as the average adult American woman. When the Mustang made its debut in 1964, the happy Ford salesmen were making deals they could hardly believe. It was full sticker price and minimum trade-in allowance for your old machine. You had two choices: take the deal or move over and let the other eight people in line have a shot at it.

The introduction of the Mustang created a textbook example of the seller's market. The size was right, the power options satisfied secretaries whose prime concern was the cost of gas, as well as the swinging singles whose concerns were looking cool and going fast.

Ford's coup stunned the automotive world. Chrysler reacted instantly by putting a fastback roof on the Valiant and calling it the Barracuda. In the Fall of 1966, Camaro became the Chevrolet entry into the lucrative market.

The end of calendar year 1966 saw a sales total of 610,212 Ponycars, with the galloping Mustang insignia on a whopping 540,802 of them. Calendar 1967 bulged out at 752,821 Ponycars, with Javelin and Firebird entering the

fray. Mustang's share that year dropped to 380,136. Camaro ran a strong second, logging 204,704, with Firebird, Barracuda and Javelin trailing in order. The world was rosy, sales were good, and the Pony market looked toward a sure million during '68.

Crystal-balling car sales has about as much science to it as forecasting race winners. The Ponies stumbled going into the clubhouse turn with total numbers for '68 slipping to 688,369. A close look at the figures shows that the decline in the popularity of the sporty set can be linked to an attendant increase in intermediate sales.

In '68, Mustang sales started to slide, as product and nameplate loyalty redrew the battle lines. Camaro increased sales by a piddling 5000, but nevertheless, it was an increase. Pontiac maintained, Javelin surged, and Barracuda took a dive. By 1969, the slide took on avalanche proportions. The introduction of a Challenger from Dodge (and accompanying rebodied 'Cuda from Plymouth) challenged hardly anybody.

By 1970 it was painfully apparent that the Age of the Ponycar was drawing to a close. Maverick began wooing the

secretaries and non-performance minded Ponycar buyers. Insurance companies slammed into the big-engined horses with crippling surcharges. The pistol shot to the head came in the form of the Pinto, Vega, and Gremlin. Ponycar sales in 1971 were down to almost half those of the peak year of '67.

By now the Ponycar picture should look like that tired old Indian drooping over his oat-burning pony, lance at the dip. Calendar '72 should have seen the end of the trail. *Should have,* but inexplicably, did not. For reasons beyond the ken of the most learned projectionists, the plunge of the Ponies slowed and very nearly bottomed out. Then there was a resurgence of sales in March and April. The true picture, however, won't be revealed until well into the '73 model year. With a damning sense of the dramatic, the troops at GM's Norwood, Ohio, plant, the facility that produces Camaros and Firebirds, went out on strike in April of '72. As the supply of GM Ponycars dried up, the demand continued until there were no more. Mustang sales picked up,

Camaro

AMX

Firebird

'Cuda

Ponycars

Javelin increased, and the Chrysler offerings retained their health.

The surviving market is apparently made up of hard-core loyalists who have been frustrated by the even higher insurance rates on Corvettes, who have become impatient waiting for a Datsun 240-Z, or who need the back seats—even if only a tiny one—for a growing family, but can still dig a sporty car. The incidence of returning Vietnam veterans with a pocketful of combat pay has added to the ranks of Ponycar drivers.

At this point, it's difficult to determine whether the Ponycar market is in that deceptive bloom of health usually seen in heart patients just before the final attack. In any event, it is maintaining a semblance of stability and refuses, for the moment, to die.

With doom in their hearts, and the bean counters snapping at their heels, the auto industry's product planners made very few changes in the offerings for '73. Most of the changes were dictated by Federal edict, and the other changes serve to refine and improve the existing product.

In late May, the Ford engineers involved in the exhaust emissions certification program got caught with their screwdrivers in the carburetor. After informing the Federal folks of their sin, Ford began the 50,000-mile test cycle from the beginning, again.

When we asked for a Mustang to test, the legal gentlemen at Ford refused to permit testing of a vehicle which may not be representative of the final product approved for sales. Their refusal was not really out of any great sense of nobility, but (more realistically) out of compliance with another Federal regulation.

We did manage to get our hands

on—and feet into—the rest of the '73 Ponycar lineup: Chevrolet's Camaro Z/28, both with and without air conditioning, Pontiac's Firebird Formula 400, the Javelin AMX, and Plymouth's snappy little Barracuda 340. Engines are of the six-liter variety with four-barrel carburetion. The exception is the Firebird, which does not offer a four-barrel version of the 350 engine.

Automatic transmissions were requested, but Chevy only had a pair of 4-speed manual-transmission cars available—so we took 'em. Beyond these little deviations, the machines were equipped as uniformly as can be obtained with the very earliest production models.

Chevrolet offers two axle ratios with the '73 Z/28—3.73 for cars without air conditioning and 3.42 for cars with. Naturally, there is a difference in performance between the two ratios, but not one you're likely to notice unless

>>>>

Camaro

AMX

Firebird

'Cuda

Ponycars

you're running on a drag strip. It took only a half-second longer to hit 30 mph from a standing start, and a full second to 60. At the end of a quarter-mile, there were only two mph and 0.5-second to tell the two apart.

Refinements on the Camaro are minimal, but very well placed. The most welcome is the addition of some real road feel in the power steering. GM uses a torsion bar actuator with the power-assist unit. By using a stiffer bar, the increased effort affords a very good idea of what's happening at the front wheels. There is a slow, 28:1 manual steering available which is satisfactory with the wide ratio tires, but is a bore to park, requiring six turns lock-to-lock.

A revised suspension for the Z/28 has a different rear stabilizer bar and

a new shock valving and comes with 15 x 7-inch wheels. Our test machine had F60-15 Firestones, which Firestone claims are a development of their racing tires. The result is a marked improvement in handling for the Z/28. Weight shift during steering reversals is quick and stable with no hunting and seeking after the fact. When you get around the corner, you're finished with it.

Unfortunately, the Camaro's instrument panel is unchanged. It still has a full bag of gauges (except for oil pressure), but for the most part they are still hidden by the driver's hands.

The Camaro's sister ship, the Firebird, waltzed out on Uniroyal GR70-15 steel-belted radial tires, bolted to the Y99 handling package. The 400-cubic-inch, 4-barrel powerplant ran through a M-40 Turbo-Hydramatic 400 transmission to a very modest 3.08:1 rear axle. Even handicapped by its long

axle ratio, the 400-Chevy incher came off almost dead even with the Chevy 350. Passing times were a couple of tenths of a second slower, but not much.

The Firebird's Y99 suspension is very much the equal to the Z/28's, but the steel-belted radials have a different effect on corners. While Camaro snapped to the stable position, Firebird took a more precise exiting technique. The radials have a bit more sidewall compliance than the Wide Ovals, but as the Firebird is pressed to the limit, its handling becomes more stable.

The Firebird has a superior instrument panel, with all of the gauges clustered in a single, visible oval. Rear deck spoiler, stripes, air conditioning and rally gauge/clock cluster are a few of the Firebird options.

Both the Firebird and Z/28 automatic transmissions come equipped with a

Pony Cars

manual shifting device that permits moving up one gear at a time by moving the shifter to the outboard detent prior to the shift, a bit like the pre-selector gearboxes of yore.

American Motors' Javelin continues with few changes, and few are needed. We found the Javelin to be a very comfortable and responsive machine. Our AMX came with a 360-cubic-inch, 4-barrel engine, automatic transmission and the standard 2.87 rear axle. Javelin gave away a couple of tenths in acceleration time all the way down the line, but still remained in the ballpark. Its handling is very positive and predictable . . . to a point. There is a tendency for the front end to plow when labored to the limit, more so than in the two GM offerings.

The ride is very comfortable, but road noise is more noticeable than in the Camaro and Firebird. As the speed increases, ride harshness gains with it.

Our adventure with Barracuda was like a visit with an old friend. The 'Cuda remains as it was, with the penalty of smog controls partially offset by the addition of an electronic ignition system. Actually, the 340-cubic-inch, 4-barrel engine appears to be little restricted by the anti-smog devices—it still comes on a bit weak until it hits 3000 rpm, and it still comes on strong as the tach hits the big numbers.

Handling is as predictable as always, and unfortunately, the ride still retains a bit too much harshness.

One department where the Ponycars shine is the fine art of bringing the whole mass to a screeching halt. Braking distances were excellent for the lot. Javelin took top honors with 32 feet for 30-0 mph, and Firebird brought up the rear with 44 feet. When we raised the ante to a stop from 60 mph, Javelin retained the crown with 155 feet and Firebird missed again at 165 feet. The Barracude managed 162 feet, with the Camaro logging 160 feet. These miniscule differences are attributable as much to tires or technique as to brakes.

Handling has improved, comfort is improved, engines are still putting out, and they all stop as quick as anything made in the U.S. Maybe that's why Ponycars are so hard to kill.   **/MT**

Camaro

AMX

Firebird

'Cuda

Camaro

AMX

Firebird

'Cuda

CHEVROLET

Camaro

A '73 newcomer to the Camaro camp is the Type LT coupe. This is a luxury-oriented touring model intended to compete with Pontiac's Esprit and has special interior and exterior appearance features, a standard 350 2-barrel, 165-hp V-8 and power steering as standard. On the inside, the LT sports luxury seat and door trims, special instrumentation and an assortment of woodgrain and silver-finished accents. Dual remote control mirrors and 14 x 7 wheels are other standard features.

The LT does not replace the Z-28, which continues for '73. This has a 255-hp version of the 350 4-barrel with hydraulic valve lifters and a new open-element air cleaner for what Chevy calls "power on demand" sound. Air conditioning is available too for the first time on the Z-28. The remaining model is the standard sport coupe.

The Camaro coupes are on a 108-inch wheelbase and measure 188.4 inches overall. That is slightly longer than last year due to the impact bumpers. You can get most any engine in Chevy's book (the 110-hp 6 being standard) with a wide variety of transmissions, most optionally column or floor shifted. Disc front brakes are standard with power assist optional. The very desirable variable-ratio power steering is another option as is a sun roof.

For those who have been annoyed by the mandatory buzzer warning systems for seat belts, the Camaro has redesigned its to sound off in forward drive only, and on floor shift manual transmissions, application of the parking brake silences the buzzer. Camaro buyers may be becoming smaller in numbers but they are still youthful, and it's pretty hard to achieve intimacy with seat belts attached, much less pursuing the romance to the tune of a strident buzzer.

The automatic floor shift has been redesigned for the '73 Camaro to incorporate a Grand Haven ratchet unit for more positive upshifting. By exerting right-hand forward pressure on the handle, the shifter unit engages the next

CAMARO Z-28

higher range in a more precise fashion. The shift handle includes a detent button at the end of the knob to prevent accidentally engaging reverse or park.

Camaros have been in very short supply throughout the '72 model season due to a prolonged strike at the single assembly plant in Norwood, Ohio. Thus, Chevrolet has no real reading on what its share of the dwindling pony car market really is in recent months. For a brief period after the current design was introduced in mid-1971, it seemed as though Camaro would pass Mustang, but the race was ended by default. Hopefully, the doubt cast by strike-lost production will not influence Chevrolet to discontinue the line, although that probability has been predicted by observers close to the scene.

CAMARO / TYPE LT

ENGINES: 250 cu. ins. (110 hp), 307 cu. ins. (130 hp), 350 cu. ins. (165, 190, 245 hp).

TRANSMISSIONS: 3-spd. manual std., 4-spd. manual opt. (350 cu. ins. only), 3-spd. auto opt.

SUSPENSION: Coil front, leaf rear.

STEERING: Manual std., variable-ratio power opt.

BRAKES: Front discs, rear drums std., power opt.

FUEL CAPACITY: 18 gals.

DIMENSIONS: Wheelbase 108.0 ins. Track 61.3 ins. front, 60.0 ins. rear. Overall length 188.4 ins., width 74.4 ins., height 49.1 ins. Weight 3205-3435 lbs. Trunk 6.4 cu. ft.

BODY STYLES: 2-dr. cpe.

CAMARO TYPE LT COUPE